42nd Biennial Meeting of the American Society of Sugar Beet Technologists 2023

Volume 60, Number 2

Savannah, Georgia, USA
27 February – 2 March 2023

ISBN: 978-1-7138-7325-9

Printed from e-media with permission by:

Curran Associates, Inc.
57 Morehouse Lane
Red Hook, NY 12571

CURRAN ASSOCIATES INC. proceedings.com

Some format issues inherent in the e-media version may also appear in this print version.

Printed by Curran Associates, Inc. (2024)

American Society of Sugar Beet Technologists
800 Grant Street, Suite 300
Denver, Colorado 80203-2944

Phone: 303-832-4460
Fax: 303-832-4468

www.bsdf-assbt.org

Additional copies of this publication are available from:

Curran Associates, Inc.
57 Morehouse Lane
Red Hook, NY 12571 USA
Phone: 845-758-0400
Fax: 845-758-2634
Email: curran@proceedings.com
Web: www.proceedings.com

42ND BIENNIAL MEETING & CONFERENCE

VOLUME 60, NUMBER 2

Abstracts

AMERICAN SOCIETY OF SUGAR BEET TECHNOLOGISTS

Hyatt Regency · Savannah, Georgia

February 27 – March 2, 2023

BIENNIAL MEETING &
CONFERENCE 2023

The American Society of Sugar Beet Technologists

PROCEEDINGS OF THE 42ND BIENNIAL MEETING, 2023

Presented at the meeting held in Savannah, Georgia
February 27 – March 2, 2023

Published by:

AMERICAN SOCIETY OF
SUGAR BEET
TECHNOLOGISTS

American Society of Sugar Beet Technologists
800 Grant Street, Suite 300
Denver, CO 80203
bsdf-assbt.org

ABOUT THE COVER

Cover photo courtesy of BSDF-ASSBT Archives. Taken at the 5th General Meeting of the American Society of Sugar Beet Technologists. The meeting was held in San Francisco, California, January 12-15, 1948.

MEETING ABSTRACTS

Table of Contents

AGRONOMY

PHYSIOLOGY, GENETICS & PLANT PESTS

PROCESSING

Abstracts of Papers Presented At The 42nd General Meeting of The American Society of Sugar Beet Technologists, DOI:10.5274/jsbr.60.2.1

BSDF-ASSBT Archives, taken in 1945

ORAL PRESENTATIONS

ABERLE, ADAM D.*[1], THOMAS J. PETERS[2], DAVID C. METTLER[3] and RYAN M. HUMANN[4], [1]North Dakota State University, Dept. 7670, PO Box 6050, Fargo, ND 58108-6050, [2]North Dakota State University and University of Minnesota, Dept. 7670, PO Box 6050, Fargo, ND 58108-6050, [3]Southern Minnesota Beet Sugar Cooperative, 83550 County Road 21, Renville, MN 56284, [4]Corteva Agriscience, 4102 Arthur Dr., Fargo, ND 58104.

Rinskor provides selective broadleaf weed control in sugarbeet.

Glyphosate-resistant (GR) waterhemp (*Amaranthus tuberculatus*) and common ragweed (*Ambrosia artemisiifolia*) are weed control challenges in sugarbeet in North Dakota and Minnesota. Sugarbeet growers control GR weeds using soil residual chloroacetamide herbicides (group 15) for waterhemp control and clopyralid (group 4) for common ragweed control. However, growers need additional effective herbicides to improve the consistency of broadleaf weed control in sugarbeet. Florpyrauxifen-benzyl (Rinskor) is a mimic auxin herbicide (group 4) which controls susceptible plants by disrupting plant growth processes. Rinskor is labeled as Loyant for postemergence grass, sedge, and broadleaf weed control in rice at 30 g ha^{-1} in Arkansas, California, Florida, Louisiana, Mississippi, Missouri, South Carolina, Tennessee, and Texas. Experiments conducted in Europe suggest sugarbeet may tolerate Rinskor to 2 g ha^{-1}. Experiments to examine crop safety and selective weed control from Rinskor were conducted at six locations in Minnesota and eastern North Dakota in 2021 and 2022. In 2021, Rinskor at 0.5, 1, 2, and 4 g ha^{-1} was applied at the 2, 6, or 10 sugarbeet leaf stage and in 2022, repeat applications of Rinskor at 0.5, 1 and 2 g ha^{-1} alone or in mixtures with glyphosate, ethofumesate and *S*-metolachlor were applied at the 2-lf and 6-lf stage. Waterhemp control in 2021 and 2022 and common ragweed control in 2022 were compared to standard sugarbeet weed control programs. In 2021, sugarbeet tolerated Rinskor at 0.5 and 1 g ha^{-1} at the 2- and 6-lf stage, but sugarbeet did not tolerate the 10-lf stage of application. In 2022, sugarbeet tolerated repeat Rinskor applications at 0.5 and 1 g ha^{-1} at the 2- and 6-lf stage. Sugarbeet root yield from Rinskor at 0.5 and 1 g ha^{-1} at the 2-lf and 6-lf stage was the same as the untreated control in 2021, but Rinskor at 1 g ha^{-1} repeated at the 2- and 6-lf stage reduced root yield in 2022. Rinskor did not affect % sucrose content in either 2021 or 2022. Rinskor alone at 0.5 or 1 g ha^{-1} only provides suppression of waterhemp or common ragweed, and will therefore need to be mixed with other sugarbeet herbicides including the chloroacetamides to achieve 95% or greater waterhemp control and clopyralid to achieve 95% or greater common ragweed control.

ANDERSON, JAY F.* and PETER J. REGITNIG, Lantic Inc., 5405 - 64th St, Taber, AB, T1G 2C4.

When do sugar beets in Alberta grow the fastest?

A sugar beet growth rate study was conducted in Alberta using a high root yield variety and a high sugar content variety. The objective of this study was to measure sugar beet root yield increases and sugar accumulation in both sugar beet varieties. Tests were conducted at 4 different locations between 2018 and 2021. Varieties were harvested approximately bi-weekly starting on July 15 and were measured for six consecutive treatment dates during the growing season ending in early October. Sugar loss to molasses was not measured due to instrumentation limitations for early harvest dates; therefore, sugar was referred to as gross sugar and not extractable sugar. For gross sugar per tonne and percent sugar, site years were analyzed separately due to an interaction that occurred between fixed and random factors. Root yield growth averaged approximately 340kgs/acre per day or 2.4 tonnes/acre per week for the high root yield variety and 310kgs/acre per day or 2.15 tonnes/acre per week for the high sugar content variety over the entire harvest period. Gross sugar per acre (GSA) increased 70kgs/acre per day or 492kgs/acre per week for the high root yield variety and 66kgs/acre per day or 460kgs/acre per week for the high sugar content variety. Gross sugar per tonne (GST) increased 1.0kgs/tonne per day or 6.9kgs/tonne per week for the high root yield variety and 1.2kgs/tonne per day or 8.1kgs/tonne per week for the high sugar content variety. Gross sugar per tonne was significantly ($P<0.05$) higher for the high root yield variety on the first harvest date, which was unexpected. On the second harvest date, the high sugar content variety had numerically higher GST. For the remaining 4 harvest dates, the high sugar content variety produced significantly ($P<0.05$) higher GST. This result provided strong evidence that the higher quality variety started slower but eventually surpassed the high root yield variety in quality. Maximum root yield and gross sugar per acre production occurred for both varieties between August 1 and August 14. Gross sugar per tonne and percent sugar accumulated the fastest between July 15 and July 31 for both varieties and slowed considerably between the second-last and last harvest dates. It was concluded that varieties vary in their sugar accumulation profiles over time and across multiple years.

BACKSCHEIDER, KELLY*[1], CRAIG ALFORD[2], RYAN HUMANN[3], DAVE JOHNSON[4] and MARISA SALAS[5], [1]Corteva Agriscience, 4849 E. 400 S. Franklin, IN 46131, [2]Corteva Agriscience, 8305 NW 62nd Ave. PO Box 7060 Johnston, IA 50131, [3]Corteva Agriscience, 4102 Arthur Dr. Fargo, ND 58104, [4]Corteva Agriscience, 2111 Dayton Ave. St. Paul, MN 55104, [5]Corteva Agriscience, 1 bis Avenue du 8 Mai 1945 78280 Guyancourt France.

Introducing Rinskor herbicide: Development of a new postemergence tool for managing weeds in sugar beet.

Rinskor active (flopyrauxifen- benzyl) belongs to the arylpicolinate class of chemistry, a new structural class of the synthetic auxin (Group 4) herbicides. Rinskor is currently registered as Loyant® herbicide in the US for use in rice where it has broad spectrum activity on grasses, broadleaves, and sedges. Rinskor has a low use rate (<30 g ai/ha), a favorable environmental and toxicology profile, rapid degradation in the soil and plant tissue, and little persistence in the environment. Due to its favorable profile, Rinskor received a "reduced-risk" review from EPA and a residue tolerance exemption in October 2019. Corteva is exploring potential opportunities to expand Rinskor into other crops including sugar beet. In Europe, Rinskor has been evaluated on sugar beet as part of a POST program where 3-4 POST applications are made in sugar beet. Results indicate that sugar beet has acceptable tolerance to Rinskor when applied from cotyledon to 6 leaves and that Rinskor provides good control of weeds in the Chenopodiacea, Apiaceae or Umbellifera families. In the United States, sugar beet growers can use glyphosate, a valuable tool in managing weeds on glyphosate-tolerant sugar beet. However, glyphosate resistance has continued to spread and additional weed control options in sugar beet are limited. Previous work has indicated that Rinskor provides excellent control of difficult-to-control weeds such as common lambsquarters (*Chenopodium album*) and other broadleaf weed species. In 2021, two trials were conducted in ND/MN utilizing Rinskor in sugar beet to evaluate crop tolerance and control of glyphosate-resistant waterhemp (*Amaranthus tuberculatus*). Initial results from this work indicated that 0.5-1.0 g ai/ha of Rinskor applied prior to 10-leaf sugar beet as the potential target rate range and application timing. In 2022, several trials were conducted throughout the sugar beet growing regions of the US. Rinskor applied alone caused visible crop response to sugar beet and the addition of glyphosate, ethofumesate, and s-metolachlor increased crop response. Two applications of a program treatment including Rinskor provided good to excellent control of common lambsquarters, waterhemp, Palmer amaranth (*Amaranthus palmeri*), kochia (*Kochia scoparia*), and common ragweed (*Ambrosia artemiisifolia*). Future work in the United States will continue to define the use rate and program treatments required to manage weeds with Rinskor in sugar beet.

BERNHARDSON, DUANE*[1], HILSCHER, ELKE[2], NARTEN, HEIKO[2] and ERICHSEN, TORBEN[2], [1]KWS Seeds LLC, 1500 S 40th St, Grand Forks, ND 58201, [2]KWS SAAT SE & Co. KGaA, Grimsehlstrabe 31, GER-3753 Einbeck.

Challenges in the transition to a new analytical process for sugarbeet quality analysis.

Sugar factories worldwide are facing several challenges, and the need to reduce costs of production and improve process efficiency is paramount. A repeatable analytical process is key to ensures consistent, high-quality data. The KWS BEETROMETER®, based on a process NIRS system, is an innovative system for standardized and full automated quality analysis of sugarbeets. The automated system reduces lab cost, lead to improvement in efficiency and low costs for maintenance and can be easily implemented in existing receiving labs. The previous measurement method involved processing sugarbeets into brei during sample preparation and then analyzing the brei for sugar content and non-sugar substances. The KWS BEETROMETER®, fully automated process, involves crumbling the beets into pieces and on a conveyor belt, a near-infrared spectrometer (NIRS) determines the sugar content as well as other quality-determining factors. As companies look to implement KWS BEETROMETER® in the receiving lab processes a big challenge is to design comparison test, based on high heterogeneity in distribution of quality compounds within and across beets. The production of a representative and reliable brei sample is influenced by the whole sample preparation process. Understanding the effects of the brei production and processing differences can aid in suitable process comparisons. Experience during implementation of the KWS BEETROMETER® at customers will be shown. During the accompanied transition from the previous measurement process to the new one many different analyses are performed. This includes not only the instrumental comparison but every process step in the quality analysis of sugar beets in the receiving labs. As an example of this, experimental results of brei processing and the influence on the reliability of quality parameter results will also be presented.

BLOOMQUIST MARK W.*[1], DIMITRI BORATYNSKI[2], JOAQUIN SANTIAGO[2], and JOHN LAMB[3]. [1]Southern Minnesota Beet Sugar Cooperative, 83550 County Road 21, Renville, MN 56284, and [2]Spreckels Sugar, 395 West Keystone Road, Brawley, CA 92227 [3]Univ. of Minnesota, 2182 Woodbridge Way, Woodbury, MN 55125.

Sprinkler irrigation versus furrow irrigation for emergence and yield in the Imperial Valley of California.

Establishing an adequate sugar beet population is one of the first challenges of sugar beet production. Sugar beet is planted in the Imperial Valley of California during September and early October. Daily high temperatures often exceed 100 degrees F in the first half of September. In addition, soils in the Imperial Valley often have significant salt content. This combination creates a harsh environment for sugar beet emergence. Fields are furrow irrigated in the Imperial Valley; however, installing sprinkler irrigation during the sugar beet emergence period is being considered by growers to increase sugar beet emergence in their fields. After the emergence period, the sprinkler pipe is removed, and the fields are furrow irrigated for the remainder of the season. Two projects were developed to quantify potential emergence and yield differences between sprinkler and furrow irrigation during the sugar beet emergence period. A field survey of sprinkler and furrow irrigated production fields planted during September was conducted by the Spreckels Sugar Agricultural Staff during the 2020-2021 and 2021-2022 growing seasons. The field survey showed a 6.3% increase in emergence with sprinkler irrigation in 2020-2021 and an 8.0% increase with sprinkler irrigation in 2021-2022 compared to furrow irrigation. Sprinkler irrigation during emergence averaged 1,119 lbs. of extractable sugar per acre greater than furrow irrigation in 2020-2021 during the first seven weeks of harvest and 312 lbs. of extractable sugar per acre greater than furrow irrigation in 2021-2022 during the first five weeks of harvest. A replicated trial was conducted at the Imperial Valley Research Center during the 2021-2022 season, looking at sprinkler versus furrow irrigation for emergence and yield. The trial was conducted as a randomized complete block with a split plot arrangement. In this trial, the sprinkler irrigated treatment emergence was 23.6% greater than the furrow irrigated treatment at the initial emergence count. The sprinkler irrigated treatment yielded 312 lbs. of extractable sugar per acre greater than the furrow irrigated treatment. The field survey and the trial at the Imperial Valley Research Center will be repeated for the 2022-2023 growing season.

CHAMPLIN, A.R.*, KWS Seeds, Inc., 5705 West Old Shakopee Road, Suite 110, Bloomington, MN 55437.

Intellectual property in the US: Fostering innovation through R&D.

Intellectual property (IP) protection in the US recognizes the value of innovation and enables purposeful access to such innovations across the sugar beet industry. US Patents and Plant Variety Protection (PVP) are two forms of IP protection that are available to continue providing breakthrough technological developments directed to supporting the continued growth and success of our industry. Research and development are essential for the continued advancement of the sugar beet industry. Research institutions, government agencies, corporations and others in the industry have made, and continue to make, significant and long-term investments in R&D to provide new innovations in sugar beet research, increasing the yield potential as well as the stability of the crop to tolerate stresses caused by disease, weeds, and insect pressure. The US IP systems serve to support these investments by rewarding innovation and incentivizing future investments in R&D. The advantages of the US IP systems are utilized throughout the sugar beet industry, ranging from innovations in crop genetics, to process technology optimization and product refinement. The United States Patent and Trademark Office (USPTO) examines patent applications under the US Patent laws directed to the utility, novelty, and non-obviousness of an innovation, in view of the caselaw interpreting those laws. In doing so, the US patent system appropriately grants a limited protection only to those innovations that meet these rigorous standards. The USPTO has further streamlined its examination process to eliminate laws of nature or natural phenomena from patent-eligible subject matter, so only useful, new, and non-obvious innovations are eligible for patent protection in the US for a statutorily defined period of time. Plant Variety Protection is another form of IP and is administered through the USDA's Plant Variety Protection Office. PVP registration protects new, distinct, uniform, and stable (DUS) plant varieties and enables the management of the use of these varieties by other breeders. A PVP owner has the ability to allow the use of a protected variety for breeding by others to develop new varieties, thereby providing a platform for continued innovation and incentivizing the development of new and improved varieties. The sugar beet industry must meaningfully address the challenges which lie ahead of us, like climate change, sustainability, and feeding the growing world population, using all the benefits of the IP systems to continue to foster innovation through R&D.

CODY GROEN*, Southern Minnesota Beet Sugar Coop, 83550 County Road 21, Renville, MN 56284.

Manipulating sugar beet plant population to compensate for high nitrogen fertility in southern Minnesota.

As the sugar content of sugar beet crops increase, profitability increases. An obstacle to raising a high sugar content sugar beet crop in the southern Minnesota growing region is high levels of total available soil nitrogen. Sugar beet yield increases with soil nitrogen, but sugar beet quality is detrimentally impacted by soil nitrogen. The objective of this study was to determine if a producer can increase harvest plant population in high nitrogen situations and mitigate the negative impacts of the excess nitrogen. This study was conducted in two environments during the 2019, two environments in 2020, and one environment in 2021 in a replicated complete block design with a split plot arrangement of six replications. Four nitrogen rates comprised the whole plot and two plant populations comprised the split plot arrangement and were applied at both locations. Fertilizer was applied by hand to reach a total available soil nitrogen level of 120, 160, 200, and 240 lbs ac^{-1} and incorporated with a field cultivator. Sugar beets were planted at a rate of 120,000 plants ac^{-1} to ensure adequate establishment, and stands were thinned to populations of 43,800 and 71,300 plants ac^{-1} 28-38 days after planting. The highest sugar content sugar beet crops were produced on soil fertilized to 120 lbs total available soil N ac^{-1}. Plant populations did not significantly impact sugar beet quality. Extractable sugar per acre was highest at 240 lbs available soil N ac^{-1} and 71,300 plants ac^{-1}. Extractable sugar per ton was highest at 120 lbs available soil N ac^{-1}.

FUGATE, KAREN K.*[1], JOHN D. EIDE[1], ABBAS M. LAFTA[2], FERNANDO L. FINGER[3], AND MOHAMED F. R. KHAN[2, 4], [1]USDA, Agricultural Research Service, ETSARC, 1616 Albrecht Blvd. N., Fargo, ND 58102, [2]Dept. of Plant Pathology, North Dakota State Univ., Fargo, ND 58108, [3]Dept. de Agronomia, Federal Univ. de Viçosa, 365712-000, Viçosa, MG, Brazil, and [4]Univ. of Minnesota Extension Service, St. Paul, MN 55108.

Transcriptomic, metabolomic, and physiological changes occurring in stored sugarbeet roots.

Sugarbeet roots are subject to tremendous physiological stress at harvest as defoliation and extraction from the soil removes roots from their source of photosynthate, water, and nutrients, and harvest and piling operations inflict injuries to roots. Roots additionally encounter environmental stress during storage from declining and fluctuating temperature and humidity conditions within piles. The physiological and environmental stresses of harvest and storage are certain to affect sugarbeet root postharvest metabolism as well as impact sugarbeet root storage properties, processing properties and sucrose yield after storage. However, the molecular changes occurring in postharvest sugarbeet roots and their relationship to root storage and processing properties are largely unknown. To better understand the molecular events occurring after harvest and their relationship to storage and processing losses, genes and metabolites that were altered in expression or concentration in roots stored at 5 and 12°C for 12, 40 and 120 days were determined and correlated to changes in root storage and processing properties. RNA sequencing and HPLC-MS analysis identified 8656 genes and 225 metabolites that were altered in expression or concentration at one or more time points during the 120 day storage period, at one or both storage temperatures. Functional annotation and pathway analysis indicated that differentially expressed genes and metabolites contributed to a diverse array of regulatory, structural and metabolic pathways, while correlation analysis identified genes and metabolites that exhibited positive or negative relationships in their expression or concentration with the deterioration in root storage and processing properties. Overall, the research indicates that a major reorientation of metabolism occurs after harvest and throughout storage and identifies genes and metabolic pathways that potentially contribute to storage losses.

KAFFKA, STEPHEN*[1], RETHWISCH, MICHAEL[2] and VAN DER MEY, BRYAN[1], [1]University of California, Department of Plant Sciences, University of California. Davis, CA 95616, [2]University of California Cooperative Extension, University of California Cooperative Extension, Riverside and Imperial Counties.

Alternatives to chlorpyrifos for sugarbeet production in the Imperial Valley of California.

This research focusses on identifying less toxic substitutes for chlorpyrifos (Lorsban®) in sugarbeet production in the Imperial Valley. Chlorpyrifos is a known neurotoxin and affects both humans and other animal and fish species. It is hazardous to applicators and workers, has been shown to accumulate in the Salton Sea and has been restricted in California. The primary insects affecting stand establishment are diverse flea beetle species, primarily pale stripped flea beetle (Systena *blanda*) and armyworm species (Spodoptera sp.). In spring with rising temperatures, army worms and diverse leaf hoppers (Empoasca sp.) can reach pest level. Low risk strategies for stand establishment emphasize IPM practices including pre-irrigation of fields, dates of planting, and seed treatments. Complementary field experiments were carried out over the 2020 to 2022 period at UC Desert Research and Extension Center (DREC), integrated with complementary trials in 6 growers' fields. Poncho Beta® (clothianidan + cyfluthrin-PB) seed treatments are a reduced toxicity standard common to all station and field experiments. Data collected include seedling emergence and loss, seedling dry weight at establishment, season-long estimates of insect pest abundance and damage, yield and root quality, and comparative costs of differing treatments. At the UC DREC, pesticide treatments were compared to untreated controls within pre-irrigated or dry-planted treatments, at two different planting dates (mid-September and mid-October). Additional soil treatments at planting are compared and include granular, soil-applied chlorpyrifos, Coragen® (chlorantaniliprole), PB seed treatments and PB plus Coragen®. These plots are split and half include additional post emergence control with Asana® (esfenvalerate), commonly used in growers' fields. A randomized block design is used in growers' fields with different treatment comparisons depending on the grower and year. Typically, three treatment comparisons were carried out using chlothianidan alone or in combination of without clothianidan. Strips (plots) 50 to 60 feet wide the length of the field were compared. The common practice used by each grower was the control in each field. Yields and pesticide use and costs were tracked for all trials and compared. The Pesticide Risk Tool (https://pesticiderisk.org/about) was used to compare treatments and risk. In general, seed treatment with clothianidin, a neonicotinid, was as effective as other pesticide combinations, tended to be less expensive, and rated lower on the pesticide risk scale compared to most other pesticide-use strategies or superior to other pesticides used in the trials. Results from the trials are reported and discussed.

KAFFKA, STEPHEN*[1], ROBINSON, PETER[2] and AKSLAND, GENE[3], [1]Department of Plant Sciences, University of California, Davis, CA 95616, [2]2Dept. Of Animal Sciences, University of California, Davis, [3]UCCE Tulare County; [4]Agronomic Services, Visalia, California.

Growing and feeding sugarbeets to dairy cows in the San Joaquin Valley of California.

Sugarbeets were widely grown throughout California during the 20th century, but the last sugar factory in the northern California closed in 2008. In the absence of a sugar industry, interest in the use of sugarbeets as a silage feedstock for dairy cows has grown, especially in the San Joaquin Valley (SJV). Increasing regulatory restrictions on the availability of water for irrigation and stricter controls on nutrient management support the use of alternative crops by dairy producers that are both water use efficient and capable of recovering both water and nutrients at depth in the soil profile. Sugarbeets can be planted in fall in the SJV, and harvested in late spring/early summer. When grown through the winter, water use is approximately half that of a summer crop (2 ac ft compared to 4 ac ft). Beets have been documented recovering water and nutrients at 3 m in depth, deeper than alternative winter annual forages. Most pests and pathogens common to summer crops are avoided during winter. If winter beet production proves viable on dairies in the San Joaquin Valley, a large market for hybrid seed would develop, potentially dwarfing previous uses for sugar alone in the region. Four trials on dairy farms growing and feeding sugarbeets for silage have been carried out during the 2018 to 2022 period. Beets were planted using strip tillage methods following corn silage in fall (late October and early November) in heavily manured fields and harvested in early to late June. Root yields varied from approximately 60 t/ac to 43 t/ac depending largely on stand establishment and uniformity. Tops were not harvested due to concerns about nitrate content and technical limitations around harvesting and feeding. No pest or disease issues were observed except for flea beetle predation on seedlings in year three at one site. A formal feeding trial was carried out in year there and beet-almond hull silage replaced corn silage and some corn grain in rations fed to high producing cows with no change in yield or milk quality. Cows consumed rations avidly and body conditions scores were equivalent throughout the trial. In years one and two, beets were co-ensiled with almond hulls and both years, in years three and four, beets were co-ensiled or ensiled with additives. Significant barrier to the wider-scale adoption of beets on diaries in the SJV remain. These focus on difficulty of preserving dry matter energy in beets during storage. In two years of measurements, shrinkage (loss) of preserved dry matter was continuous throughout the storage period and averaged 35 %, compared to 5% or less for corn silage. Other barriers to adoption are the need for beet harvesters, some adjustments in forge wagons, the need for ag bags compared to bunker silos for preservation, and slower harvests and handling. These problems have not been solved adequately to date, especially those concerned with storage DM losses.

KANDEL, SHYAM*, USDA-ARS, 1307 18th St N, Fargo ND 58102-2765.

Study of storage pathogens and bacterial contamination in sugarbeet processing facilities.

Sugarbeet is one of two crops supporting sugar demand in the U.S. In the Red River Valley of North Dakota and Minnesota, postharvest sugarbeet roots are stored for up to eight months prior to processing into sugar. During storage, endogenous metabolism and microbial activity cause root deterioration and sucrose loss. Furthermore, microorganisms, especially bacteria present in the processing streams at sugar factories cause additional sucrose loss and processing challenges due to biofilm formation and exopolysaccharide production. We recently examined postharvest sugarbeet roots to identify the major microbial isolates that are present and likely to cause disease and sucrose loss during storage. Bacteria present in factory diffusion tower raw juice were also identified. Knowledge of the microbial communities present in postharvest sugarbeet roots and factory processing streams aid in developing management strategies that limit sucrose loss during sugarbeet root storage and processing.

KOCH, HEINZ-JOSEF*[1], GRUNWALD, DENNIS[1], ESSICH, LISA[2], KUEHLING, INSA[3], RUSER, REINER[2] and KAGE, HENNING[3], [1]Department of Agronomy, Institute of Sugar Beet Research, Holtenser Landstr. 77, 37079 Goettingen, Germany, [2]Department of Fertilization and Soil Matter Dynamics, Institute of Crop Science, University of Hohenheim, Fruwirthstr. 20 70599 Stuttgart, Germany, [3]Department of Agronomy, Institute of Crop Science and Plant Breeding, University of Kiel, Hermann-Rodewald-Str. 9, 24118 Kiel, Germany.

Cover crop N effects are different for subsequent sugar beet and maize.

Cover crops (cc) are supposed to decrease the soil mineral N content (Nmin) during winter and increase the N supply to subsequent main crops due to mineralisation of N previously prevented from leaching. However, data on N supply from cc grown before sugar beet and maize have rarely been reported for Central European conditions. Therefore, our study aimed to provide information for cc differing in frost resistance and biomass quantity. In 2018/19 and 2019/20, field trials were conducted on three Luvisol sites in Germany, comprising a sequence of either fallow or autumn sown cc (oil radish, saia oat, spring vetch and winter rye) followed by sugar beet and maize main crops sown next spring. Immediately before sowing, cc plant material was incorporated by tillage to

10-25 cm soil depth. The N effect of cc on sugar beet and maize was calculated as the difference in plant N uptake after a specific cc compared to bare fallow for distinct phases of the growing season. Winter rye and oil radish cc revealed the greatest potential for scavenging nitrate from the soil profile while reductions caused by frost-sensitive saia oat and spring vetch were more variable. The amount of N in the cc biomass was negatively correlated with soil Nmin in autumn. Thus, for environmentally effective cover cropping in Central Europe, species with a sufficiently high frost tolerance should be chosen. Despite cc N uptake of up to 170 kg N ha^{-1} and C:N ratios were < 20, for sugar beet a positive N effect was found between March and July only and amounted to 20 kg N ha^{-1} at maximum, while between August and September, net immobilization was predominant with up to 40 kg N ha^{-1}. In contrast, for silage maize the N effect was close to zero from sowing until July but clearly positive with values up to 60 kg N ha^{-1} in the second half of the growing season. Differences among cc species were not consistent across the site/years investigated. Sugar yield was lowest after rye and correlated positively with Nmin in spring. Maize total plant dry matter yield revealed no clear differences among cc. Correlation between yield and cc N effect was low and not improved by a multiple regression approach. Thus, factors other than in-season N supply from cc apparently impacted yield to a larger extent. For sugar beet, we hypothesize that the rapid depletion of the upper 30 cm soil layer from nitrate by the crop might cause substantial N immobilization during cc biomass decomposition in the summer months. In contrast, under maize N uptake is delayed, fostering microbial breakdown of and N release from the cc biomass.

KOCH, HEINZ-JOSEF*[1], TINO KRESZIES[2], NORBERT CLAASSEN[2], HENRIK FUELLGRABE[2] and KLAUS DITTERT[2], [1]Dept. of Agronomy, Institute of Sugar Beet Research, Holtenser Landstr. 77, 37079 Goettingen, Germany, [2]Plant Nutrition & Crop Physiology, Dep. Crop Sciences, Univ. Goettingen, Carl-Sprengel-Weg 1, 37075 Goettingen, Germany.

Optimum potassium supply increases early sugar beet growth and autumn yield.

Adequately high potassium (K) nutrition is essential for high sugar yield and quality of beet crops.: We quantified the effect of potassium deficiency on sugar beet dry matter and sugar yield formation during the growing season. Sugar beets were grown on low, medium, and high soil K concentrations in a long-term K fertilizer field experiment on a silt loam alluvial soil. Plants were harvested at four time points during the growing season, including the final harvest. At each harvest, the dry matter yield and K concentration of leaves and beets (when applicable) were recorded. In autumn, sugar concentration and beet quality were measured. Low soil K concentrations resulted in an approximately 10% lower sugar yield compared to high soil K supply. Reduced growth of low K plants already occurred during the germination phase and in the first weeks thereafter, whereas later growth rates were similar. Higher soil K levels increased K uptake rate and thus a higher final total K uptake. High K plants showed rapid early growth and maximum growth rates were achieved earlier than by low K plants. The initial differentiation between the low and high K treatments persisted due to similar growth rates in subsequent phases at both low and high soil K. It is concluded that under overall low soil K conditions, adequate amounts of K fertilizer should be applied to meet the demand of sugar beet seedlings for maximum growth and final sugar yield. Banded K fertilizer application next to the plant rows is supposed to be efficient for this purpose.

LAMB, JOHN A.*[1] and MARK W. BLOOMQUIST[2]. [1]Univ. of Minnesota, 2182 Woodbridge Way, Woodbury, MN 55125, and [2]Southern Minnesota Beet Sugar Cooperative, 83550 County Road 21, Renville, MN 56284.

Nitrogen application rate and timing in the Imperial Valley of California.

Growers in the Imperial Valley have adopted glyphosate resistant varieties in their sugar beet production system. One of the advantages of the use of glyphosate resistant varieties is the reduction of the need to cultivate for weed control at layby (November). This cultivation operation at layby was also combined with a split application of nitrogen. The cultivation at layby required that irrigation basins be deconstructed for equipment access to the field and then rebuilt following the cultivation. With the advent of glyphosate resistant varieties, weed control is obtained without cultivating and thus the irrigation basins do not need to be deconstructed. This requires growers to apply their nitrogen fertilizer pre-plant instead of the former split application. The objective of this study was to determine the effect of nitrogen rate and timing on sugar beet root yield and quality. The study was established at four locations from 2017 to 2020. The treatments were a factorial combination of eight nitrogen application rates (0, 40, 80, 120, 160, 200, 240, and 280 lbs. N per acre) and two application timings (pre-plant and layby). This study suggests that delaying N application until November is not needed and that the optimum N application rate for sugar beets harvested in June and July was 220 lbs. N per acre with another 100 lbs. soil nitrate-N per acre in the surface 43 inches of soil at planting.

LYSTAD, ALEXA L.*[1] and THOMAS J. PETERS[2], [1]North Dakota State University, Dept. 7670, PO Box 6050, Fargo, ND 58108-6050, [2]North Dakota State University and University of Minnesota, Dept. 7670, PO Box 6050, Fargo, ND 58108-6050. **Reinventing common ragweed control with clopyralid in sugarbeet.**

Common ragweed (*Ambrosia artimisiifolia*) is a troublesome summer annual broadleaf weed in sugarbeet in Minnesota and North Dakota. Growers attending the 2022 sugarbeet growers' seminars reported common ragweed as their second most troublesome weed following waterhemp (*Amaranthus tuberculatus*). Past experiments investigating chemical control options reported targeting common ragweed less than 5-cm with repeat glyphosate plus clopyralid applications at 1,010 g ha^{-1} plus 53 g ha^{-1}, respectively, provided 92% control. Repeat applications of clopyralid plus glyphosate were more effective on both small (≤5 cm) and larger (≤10 cm) common ragweed; however, common ragweed 15-cm or greater were too large for POST control in sugarbeet. Recent greenhouse evaluation of common ragweed sourced from fields with weed control failures confirmed that the application of glyphosate alone is no longer an effective mode of action for common ragweed control. In addition, certain common ragweed populations from 2021 also demonstrated alarming tolerance to clopyralid; however, clopyralid eventually provided common ragweed suppression at 158 g ha^{-1}. A field experiment conducted in 2022 considered common ragweed control from a more concentrated formulation of clopyralid (Stinger HL) at different rates, timings, and tank mixtures. Results suggest common ragweed populations are adapting to sugarbeet herbicides and sizes ranging from 5- to 10-cm are becoming increasingly difficult to control. Clopyralid rates to 105 g ha^{-1} plus glyphosate provided unacceptable control when applied to common ragweed greater than 5-cm. Targeting common ragweed less than 5-cm with repeat clopyralid applications at 79 g ha^{-1} or greater plus glyphosate at 1,050 g ha^{-1} provides the greatest common ragweed control in sugarbeet. Timing herbicide applications to common ragweed size rather than sugarbeet growth stage is crucial.

METTLER, DAVID C.*, JOHN A. LAMB and MARK W. BLOOMQUIST, Southern Minnesota Beet Sugar Cooperative, 83550 County Road 21, Renville, MN 56284.

Soil fertility management for *Zea mays* grown after unharvested *Beta vulgaris* in southern Minnesota.

The goal of Southern Minnesota Beet Sugar Cooperative is to optimize the factory's capacity. To do this the grower's goal is to raise enough high-quality sugar beets to meet the needs of the factory. This may occasionally mean that some sugar beet acres will not be harvested because of greater than anticipated root yield and a limited factory slice capacity. Little information exists on management practices for optimum field corn production following unharvested sugar beet. The objectives of this study were to determine if the unharvested beet roots need to be removed, if starter fertilizer would increase corn grain production, and if corn needs more nitrogen applied after unharvested beet roots compared to harvested beet roots. This trial was conducted as a randomized complete block with four replications and repeated over four years. Plots contained either harvested or unharvested beet roots from the previous growing season with all the fertilizer being applied to the corn in the spring based upon a two-foot soil sample.

All four years had very different environmental conditions. There was an interaction between treatment and year for corn grain yield. However, this interaction was in the magnitude of grain yield response. In all years, corn grown in plots where beet roots were not harvested had less grain yield than corn grown in plots where the beet roots were harvested. In all production years, the use of an extra 40 lbs of nitrogen per acre over the standard recommendation increased the corn grain yield in the plots where the beet roots were not harvested. The results from all years shows the importance of additional nitrogen for field corn production after unharvested beet roots because of the tie-up of nitrogen from the extra carbon left in the soil by the unharvested beet roots.

OLSEN, DAVEY J*, Sugarbeet Quality Improvement (SBQI), Amalgamated Sugar, 1951 S Saturn Way, Suite 100, Boise, ID 83709.

Sugarbeet variety selection for high nitrogen soils.

Sugarbeet production on soils with a long history of manure or dairy compost additions can be challenging. Often containing more than 400 lb N/acre, these soils are much higher in nitrogen than is recommended for optimal sugarbeet productivity. High nitrogen is known to reduce sugar % and increase impurities in sugar juice. This results in a higher sugar purity penalty which then reduces the beet payment to growers. In addition to manured fields, high nitrogen can also be a problem for sugarbeets following onion or potato rotations, which are often highly fertilized. Unable to reduce soil nitrogen, growers have few options available to them. Selecting a variety that performs well under these conditions therefore becomes critical. While there is some anecdotal evidence suggesting certain varieties do perform better than others under high nitrogen conditions, these claims have not been tested. This study reports the findings of Agridata analysis and field trial research investigating the performance of 30 commercial sugarbeet varieties under normal (250 lb N/A) and high (450 lb N/A) nitrogen growing conditions. Although most varieties showed increased yield, overwhelmingly, high-N led to reduced sugar %, higher conductivity, and greater sugar juice impurities in the form of higher brei nitrates. Despite this, there were clearly some varieties that performed better than others and were able to overcome their reduced sugar % and purity penalties by delivering much higher yields. This work provides variety recommendations for growers with high nitrogen soils.

OLSEN, DAVEY J*, Sugarbeet Quality Improvement (SBQI), Amalgamated Sugar, 1951 S Saturn Way, Suite 100, Boise, ID 83709.

Phosphorus deficiency symptoms in Idaho sugarbeet.

Phosphorus (P) is second only to nitrogen in importance for optimizing growth, crop yields, and quality. Despite its importance deficiency in sugarbeets can be common, particularly during early spring when soils are cold and sugarbeet roots are small. Diagnosis is not always easy as phosphorus deficiency is widely regarded as the most difficult nutrient disorder to diagnose in sugarbeets. It is not uncommon for P-deficiency to be misdiagnosed as nitrogen (N) deficiency, resulting in potential for over-supply of N and reduced beet quality. A wide range of symptoms are attributed to P-deficiency, many of which, such as purpling of leaves, are often listed in one diagnostic key only to be omitted altogether by another. This has led to disagreement amongst sugarbeet agronomist as to what visual symptoms best describe P-deficiency in sugarbeet grown on Idaho soils. In response to this, a glasshouse study was established in 2020 to document the visual symptoms of P-deficiency across 4 levels of soil P supply (3, 10, 15, and 27 ppm-P). A follow up study, with the same treatment levels, was established in 2021 to investigate symptom expression across 10 commercial sugarbeet varieties. These studies found that many commonly described P deficiency symptoms, such as delayed emergence, yellow cotyledons, purpling of leaves, brown veining of older leaves, crinkling of leaf margins, and leaf cupping, were not reliable indicators of P-deficiency. Rather, the most reliable and consistent indicators of P-deficiency were slow leaf production and reduced plant size, smaller and somewhat darker green first true leaves, and some degree of upright leaf habit in young leaves. All ten varieties showed remarkably similar symptom expression. This work has helped Amalgamated crop advisors better recognize the symptoms of sugarbeet P deficiency in Idaho soils.

OLSEN, DAVEY J*, Sugarbeet Quality Improvement, Amalgamated Sugar, 1951 S Saturn Way, Suite 100, Boise, ID 83709.

The impact of soil sampling timing on soil test results.

An accurate soil test provides the foundation for determining the optimal amount of fertilizer required for a sugarbeet crop. While some nutrients do not change much from fall to spring, others are more variable. It is well understood that available soil nitrogen (N) is dynamic and changes over time in response to natural processes such as mineralization, leaching, denitrification, immobilization, and human interventions such as tillage and residue management. For this reason, it is commonly recommended that soil sampling for N occur as close to planting as possible. Soils sampled too far ahead of planting are less able to accurately measure soil N available to sugarbeets at planting. This results in N fertilizer recommendations that may be sub-optimal for sugarbeet production and/or profitability. Despite these concerns, sugarbeet soils are routinely sampled in the fall, often as early as August. In the lead up to the 2020 crop year, Amalgamated Sugar sampled 1173 fields prior to January. This represented 43% of total contracted fields. With more than 2000 sugarbeet fields to sample each year, it would be logistically challenging to require that all fields be spring sampled. Instead, it is possible that improvements could be made to the fall soil sampling program if it was understand how much soil test N varies due to time of sampling. This report details the results of monthly soil sampling at 15 commercial sugarbeet fields in the fall of 2021 and spring of 2022. We found that soil samples taken in the fall significantly underestimated total available N relative to spring samples. As a result, fertility recommendations based on fall soil tests overestimated fertilizer N requirement resulting in unnecessary cost to the grower and the potential for negative impacts on sugar % and sugar quality. This work suggests that fertilizer recommendations based on fall soil tests could be improved by considering their tendency to underestimate soil available N relative to spring soil tests.

OLSEN, DAVEY J*[1], and DAVID D. TARKALSON[2], [1]Sugarbeet Quality Improvement, Amalgamated Sugar, 1951 S Saturn Way, Suite 100, Boise, ID 83709 and [2]USDA-ARS Northwest Soils and Irrigation Research Laboratory, Kimberly, ID 83341.

Static Nitrogen: Evaluating an alternative method for determining nitrogen requirement in sugarbeet.

The current method for determining nitrogen (N) requirement in sugarbeet, the Yield-Based Approach, calculates N fertilizer requirement by applying a 'lb N per ton beets' multiplier to the projected yield for a given field. As yields continue to increase over time, this method results in higher and higher N recommendations and a constant demand for research-based updating of the N multiplier. Despite the trend for increasing yields, research has shown the amount of N required to achieve those yields has remained relatively stable. For this reason, researchers from Amalgamated Sugar's SBQI team, in partnership with soil scientists at the USDA-ARS, have been exploring possible alternatives to the yield-based approach. From 2014 to 2018 a study of 14 fertility field trials concluded that nitrogen requirement could be reduced in the Pacific Northwest and that a flat-rate, or static, approach should be considered as an alternative to the yield-based approach (Tarkalson et. al, 2014). Further work in 2018-19 at 6 field sites identified a static-N range of 180–230 lb N/acre was optimal to maximize sugarbeet yield and quality. This report details the findings of static N research at a further 6 field sites in 2020 and 2021. These studies compared the Yield-Based approach to a range of Static N alternatives (180, 205, 230 lb N/A), as well as N recommendations from a commercial soil testing laboratory. Results showed the Static rate approach better matched N supply with crop need, and that both the Yield-Based Approach and Soil Lab recommendations significantly over-estimated N fertilizer requirement. The current Yield-Based method resulted in an average of 81 lb N/acre excess fertilizer N when compared to the Static N approach.

PETERS, THOMAS J.*[1], ALEXA L. LYSTAD[1], DAVID C. METTLER[2] and EMMA L. BURT[3], [1]North Dakota State University and University of Minnesota, NDSU, Dept. 7670, PO Box 6050, Fargo, ND 58108-6050, [2]Southern Minnesota Beet Sugar Cooperative, 83550 County Road 21, Renville, MN 56284, [3]Minn-Dak Farmers Cooperative, 7525 Red River Rd., Wahpeton, ND 58075.

Dicamba and glufosinate integrated with soil residual and POST herbicides improves control of GR weeds in sugarbeet.

Glyphosate resistant (GR) common ragweed (*Ambrosia artimisiifolia*), kochia (*Bassia scoparia*), and waterhemp (*Amaranthus tuberculatus*) are weed control challenges in sugarbeet. Sugarbeet producers control GR common ragweed with clopyralid at rates ranging from 53 to 105 g ha^{-1} and GR waterhemp using soil residual herbicides applied after planting or between 2-lf and 8-lf sugarbeet stage. Timely rainfall to incorporate soil residual herbicides was absent in 2021 and 2022, complicating waterhemp control. GR kochia control in sugarbeet is best when effective herbicides are used in the crop sequence preceding sugarbeet. Ethofumesate is the most reliable sugarbeet herbicide for kochia control but is dependent on rainfall for incorporation. Dicamba and glufosinate integrated into the weed control program with Truvera® sugarbeet will benefit sugarbeet producers. Field experiments were conducted in Minnesota and North Dakota in 2021 and 2022 to evaluate sugarbeet tolerance and to redefine a weed control program to control GR weeds. In tolerance experiments, dicamba at 0.56 kg ha^{-1} plus a volatility reduction adjuvant and a drift reduction agent at 1.46 L ha^{-1} and 0.5% v/v, respectively, at 2- to 4-lf sugarbeet caused stature reduction injury 1 to 7 and 8 to 14 days after application as compared with the glyphosate control in 2021. However, sugarbeet injury was negligible from dicamba 15 or more days after application or from glufosinate at 0.62 kg ha^{-1} across application timing. Root yield from dicamba applied at the 10- to 12-lf stage was less than root yield from the glyphosate control or from dicamba PRE, dicamba at the 2- to 4-lf or the 6- to 8-lf stage and from glufosinate at the 2- to 4-, 6- to 8 or 10- to 12-lf stage. Dicamba or glufosinate did not affect percent sucrose compared with the glyphosate control. Dicamba alone or dicamba mixed with ethofumesate PRE controlled early emerging waterhemp better than and dicamba and glufosinate with glyphosate, and chloroacetamide herbicide(s) provided season long waterhemp control better than our best currently commercialized waterhemp control programs in 2021 and 2022. Treatments combining repeat clopyralid applications at 79 g ha^{-1} with glufosinate or glyphosate or dicamba PRE followed by repeat applications of clopyralid at 79 g ha^{-1} with glufosinate or glyphosate provided burndown control of up to 5-cm weeds and provided greater than 95% season-long common ragweed control or greater common ragweed control than repeat glyphosate and clopyralid applications at 1.05 kg ha^{-1} and 79 g ha^{-1} with NIS and AMS in 2022. Repeat glyphosate applications with dicamba or glufosinate at the 4 to 6-lf or 8- to 10- lf sugarbeet stage, respectively, provided greater than 95% kochia control of up to 10-cm kochia as compared with kochia control from repeat glyphosate applications alone or repeat glyphosate and ethofumesate applications following ethofumesate PRE.

REGITNIG, PETER J.* and JAY F. ANDERSON, Lantic Inc., 5405 - 64th Street, Taber, Alberta, T1G 2C4.

Nitrogen and irrigation effects on sugar beet emergence.

Every year a portion of Alberta growers apply urea nitrogen fertilizer in spring prior to seeding sugar beets. Twenty-two separate trials over a 5-year period between 2015 and 2020 were conducted to assess the potential for sugar beet stand loss when urea is soil incorporated prior to planting and irrigated prior to sugar beet emergence. In addition, nine separate trials were conducted between 2015 and 2021 to assess the potential for sugar beet stand loss when urea is surface applied and irrigated after planting but prior to emergence. Urea rates up to 200 lbs N/acre were applied in all trials, with the 200 lb rate being used to accentuate and differentiate treatment effects. When urea was soil incorporated, highly significant ($p<0.01$) sugar beet stand reductions averaging 28% were observed with applications of 200 lbs N/acre urea in 6 tests where a single irrigation ≤0.5 inch was applied. Where a single irrigation ≥0.6 inch was applied, sugar beet stand reductions with 200 lbs N/acre averaged 11% and were only significant ($p<0.05$) in 2 of 6 trials. Stand reductions with 100 lb N/acre soil incorporated urea averaged 12% and were significant ($p<0.05$ and $p<0.01$) in 4 of 6 trials where a single irrigation of ≤0.5 inch was applied. When a single irrigation ≥0.6 inch was applied, sugar beet stand reductions with 100 lbs N/acre averaged 2% and were not significant in any trials. Results were similar for single and multiple irrigations with 100 and 200 lbs N/acre soil incorporated applications. When urea was surface applied and irrigated prior to sugar beet emergence, sugar beet stand reduction ranged from 2 to 98% with applications of 200 lbs N/acre, with higher stand reductions observed for lower irrigation plus rainfall volumes. Results from these trials show that greater amounts of applied water can mitigate the toxic effects of spring urea applications. Many Alberta growers use lower irrigation volumes when watering for sugar beet emergence. When spring urea is incorporated at 100 lbs N/acre with current watering practices, a sugar beet stand loss in the order of 12% could be expected.

STEINKE, KURT* and ANDREW CHOMAS, Michigan State University, Dept. of Plant, Soil, and Microbial Sciences, East Lansing, MI 48824.

Changing fertilizer strategies to address new challenges in sugarbeet management: Early harvest, varietal differences, and sugar quality.

Nutrient management strategies have always been a focal point of sugarbeet production. However, with the adoption of the 30-300 Initiative (i.e., 30 T A^{-1} and 300 lbs sugar T^{-1}), Michigan sugarbeet production continues to emphasize improving beet quality as compared to yield. New obstacles continue to impact agronomic management including a more variable climate, early harvest intervals, updated varietal characteristics, and managing an N-responsive cropping system that growers may or may not wish to respond to N applications. Although management has mostly evolved to being site- or field-specific, little work has been done investigating how nutrient management may change based on sugarbeet varietal characteristics. More defensive varieties with greater disease tolerance may respond differently compared to more aggressive varieties with greater tonnage. Field studies were established to evaluate sugarbeet varietal response (defensive vs. aggressive) to specific fertilizer management strategies and early vs. conventional harvest intervals. The study was blocked by two harvest timings (early and conventional) and two varieties (C-G675 and C-G919). All treatments received 60 lb. N A^{-1} at planting applied 2×2. Four fertilizer strategies consisted of only 60 lb. N A^{-1} applied 2×2 at-plant, 60 lb. N A^{-1} applied 2×2 and 100 lbs. N A^{-1} sidedress coulter inject at 4-leaf stage, 60 lb. N A^{-1} applied 2×2 and 100 lbs. K_2O A^{-1} surface applied next to row at canopy closure (~20 leaf stage), and 60 lb. N A^{-1} applied 2×2 along with 100 lbs. N A^{-1} sidedress coulter inject at 4-leaf stage and 100 lbs. K_2O A^{-1} surface applied next to row at canopy closure (~20 leaf stage). Climate variability continued to impact growth and production in 2021 with moist autumn conditions decreasing sugar concentrations by 0.37% in October harvest as compared to August. Despite a mean yield increase of 16 T A^{-1} with conventional as compared to early harvest, no interaction between variety and harvest timing occurred in 2021. Starter N was sufficient for early harvest specifically in 2021 with no tonnage differences observed compared to starter with sidedress N. Field results from 2022 will be included after harvest is complete and compared alongside 2021 field data.

SPRAGUE, CHRISTY L.*[1], Department of Plant, Soil, and Microbial Sciences, Michigan State University, 1066 Bogue Street, East Lansing, MI 48824.

Michigan State University's experiences managing glyphosate-resistant waterhemp in sugarbeet.

Glyphosate-resistant waterhemp is quickly becoming one of the most difficult weeds to manage in sugarbeet. Extended emergence, rapid growth rates, and the lack of postemergence herbicide options necessitate the use of creative options to manage waterhemp. Over the past six years we have examined various strategies and tested various herbicides for waterhemp control and sugarbeet tolerance. Acifluorfen (Ultra Blazer), phenmedipham (Spin-Aid), phenmedipham + desmedipham (Betamix), and the residual herbicides metamitron, ethofumesate and several Group 15 herbicides have all been components of this research. To date, the most consistent waterhemp control includes the use of a preemergence herbicide either ethofumesate (1.12 kg ha^{-1}) or s-metolachlor (0.53 kg ha^{-1}) followed by one of the overlapping herbicide programs of two applications of: 1) s-metolachlor (Dual Magnum) at 1.06 kg ha^{-1}, 2) acetochlor (Warrant) at 1.26 kg ha^{-1}, or 3) dimethenamid-P (Outlook) at 0.56 kg ha^{-1} applied at 2- and 6-8 leaf sugarbeet. In most cases, these programs also did not significantly influence sugarbeet yield. It is important to remember that the residual herbicide will not control emerged waterhemp. This is where understanding the time of waterhemp emergence in relation to sugarbeet planting is important. Most of Michigan's waterhemp populations start to emerge the 3rd or 4th week of May. The first application of a residual herbicide would need to be applied prior to waterhemp emergence. If sugarbeets are planted later, the use of a preemergence herbicide would be needed.

TARKALSON, DAVID*[1], DAVEY OLSEN[2] and DAVE BJORNEBERG[1], [1]USDA-ARS, 3793 N 3600 E, Kimberly, ID 83341, [2]Amalgamated Sugar Company, 1951 S Saturn Way, Suite 100, Boise, ID 83709.

Economic value of nutrients in sugarbeet processing precipitated calcium carbonate.

Precipitated calcium carbonate lime is a byproduct of sucrose extraction at sugarbeet processing factories in Idaho. Each year 351,000 Mg of this lime is produced and stockpiled at sugarbeet factories. There is currently no viable disposal strategy for this material and these piles continue to grow in size each year. However, with an average total P content in PCC of 150 kg P_2O_5/Mg and high fertilizer P prices, the application of PCC as a P source can be a practical alternative to fertilizers.

TARKALSON, DAVID*[1], BJORNEBERG, DAVE[1], NEHER, OLIVER[2], OLSEN, DAVEY[2] and DEAN, GREG[2], [1]USDA-ARS, 3793 N 3600 E, Kimberly, ID 83341, [2]Amalgamated Sugar Company, 1951 S Saturn Way, Suite 100, Boise, ID 83709.

Effects of sugarbeet processing precipitated calcium carbonate on crop production and soil properties.

Precipitated calcium carbonate lime is a byproduct of sucrose extraction at sugarbeet processing factories in Idaho. Each year 351,000 Mg of this lime is produced and stockpiled at sugarbeet factories. There is currently no viable disposal strategy for this material and these piles continue to grow in size each year. The simplest solution would be to apply this PCC directly to agricultural fields each year, however the effects of PCC on high pH soils and southern Idaho crop rotations are not well understood. Due to high amounts of lime production (351,000 Mg $year^{-1}$), dA study was conducted at the USDA-ARS laboratory in Kimberly, ID to determine the effects of PCC application to an alkaline silt loam soil on sugarbeet, dry bean and barley production and soil properties. Three PCC treatments (rate and timing) and an untreated control were compared. The PCC had no effects on crop production factors and most soil properties. The only significant effect of PCC treatments was an increase in soil phosphorus (P) concentrations compared to the control. The PCC can serve as a P fertilizer. For all crops in this study, PCC was applied at rates that resulted in applied P levels that were 1.6 to 5.3 times greater than even the highest published recommended P rates. Compared to the control, bicarbonate soil P concentrations increased by 25% and 73% for the final PCC application amounts of 26.9 kg ha^{-1} (6.7A treatment) and 89.7 kg ha^{-1} (6.7A and 89.7T treatments), respectively. The PCC used in this study can safely be applied (at rates up to 87.9 kg ha^{-1}) to heavier textured alkaline soils in the local growing area. Disposing of PCC in this way is a viable strategy for reducing PCC stockpiles.

TSCHOEP, HENDRIK*, EYLAND DAVID, KRISTOF GOVAERTS, AND JEROME DE PESSEMIER, SESVanderHave, Industriepark 15, 3300 Tienen, Belgium.

Mitigating drought impact through variety improvement.

Drought is one of the main constraints in sugar beet cultivation. Empirical studies and model simulations demonstrated yield reductions up to 40%, resulting in a significant financial impact on the beet sugar industry. The impact of water shortage is expected to aggravate in the future, as the frequency and intensity of drought spells are expected to increase with climate change. Therefore, yield stability under water-deficit conditions is becoming a crucial trait in sugar beet breeding. By combining continuous environmental and satellite data we aim to classify sugar beet growing environments along different drought intensities. Based on these classifications, genetic diversity in drought tolerance across multi-location trial networks can be assessed. Furthermore, we discuss the possibilities and the approach that we employ for further variety improvement for drought tolerance. □

POSTER PRESENTATIONS

AGHAEI, MAJID*[1], MASOUD HONARVAR[2] and MOHSEN BAZRAFSHAN[3], [1]Marvdasht Sugar Beet Factory, [2]Islamic Azad University, Science and Research Branch, Tehran, Iran, 1477893855, [3]Sugar Beet Research Department, Fars Agricultural and Natural Resource Research and Education Center, AREEO, Shiraz, Iran.

Changing the sugar beet cropping system in Iran, from spring to autumn in order to increase water productivity.

Sugar beet is one of the industrial products of the agricultural sector that plays an important role in providing sugar to the household food basket in Iran. Sugar beet production has a history more than one century in our country; Iran is located in semi-arid area with 250 mm annual precipitation. Climatic conditions of Iran are in a manner that sugar beet can be cultivated in either spring planting or autumn planting. All spring sugar beet cultivation should be irrigated. Water is already the most limiting factor in sugar beet cultivation; it is necessary to take decreased water consumption and increased water productivity in sugar beet production. Autumn planting of sugar beet as compared to its spring cultivation demands less application of water in semi-arid region such as Iran. The water use efficiency for spring and autumn sugar beet cultivation is about 600 and 1000 gram of sugar per m^3 of water used respectively. Therefore, cultivation of sugar beet in autumn is considered as a better choice to be benefited from autumn and winter precipitation and escaping from water deficit crises. Autumn cultivation of sugar beet is one of the strategies for increasing sugar beet production with minimum amount of water consumed due to the use of autumn and winter precipitation and avoiding the need for irrigation in very dry summer air. According to sugar beet syndicate annual report (2021, 2020, 2019, 2018), the total autumn sugar beet cultivation were 26451, 16280, 12054, 15028 ha respectively. The average of sugar beet yield was 53, 57, 55, 65 ton/ha and sugar content was 14.74%, 14.85%, 13.89% and 14.57% respectively. Autumn sugar beet cultivation began since 1963 in Khozestan province and now due to the change in the cropping system it has spread to other provinces like Fars, Golestan, Kermanshah and Ilam. Considering the unique climatic conditions of Iran, development of autumn sugar beet cultivation will be caused a revolution in agriculture, help increase employment in agriculture and industry, considerably more produced white sugar and enhancement of water productivity. Therefore, the sustainability of this strategic product is of great importance for ensuring its supply and the partial sustainability of Iran's food security.

ALDER, CLARKE*[1] and ALBERT ADJESIWOR[2], [1]Amalgamated Sugar Company, 1951 South Saturn Way, Suite 100, Boise, ID 83709, [2]University of Idaho, 3806 North 3600 East, Kimberly, ID 83341.

Herbicide carryover in sugarbeet - an assessment.

Chemical carryover from the previous crop into sugarbeet is a common problem that crop consultants and growers face each year in southern Idaho. Crops such as dry beans, potatoes, onions, and others allow chemicals to be applied that may be detrimental to sugarbeet if applied at the incorrect timing or if sugarbeet is planted earlier than is recommended by the guidelines found on the pesticide label. Often, time is of the essence when making management decisions

once a problem of this nature is manifest in late spring. This study was designed to help growers and crop consultants make management decisions by assessing potential yield losses and other potentially negative effects several chemicals found in the rotation may have on the current sugarbeet crop under both tilled and no-tilled ground. Chemicals were applied in late spring of the previous year to bare ground and the ground was managed as if a crop existed for the remainder of the year. Tillage treatments were administered in late fall and sugarbeets were planted into the existing plots the following spring. Preliminary data show no significant differences between nitrates, percent sugar, or predicted thick juice purity between any of the treatments. The largest significant factor was yield for which imazamox, terbacil, and metribuzin exhibited the lowest yield in both tilled and no-tilled plots. Consequently, pounds of estimated recoverable sucrose per acre were also affected. No-till plots had significantly higher nitrates and conductivity and were slightly lower in pounds of estimated recoverable sucrose per ton and predicted thick juice purity than tilled plots. Losses in yield can be attributed to several factors including root structure, stand establishment and loss, and growth rate of the crop. Sugarbeets will be planted into the same area in 2023 to assess possible further damage from longer persisting chemistries. Results from this study including data, pictures, and symptom descriptions will help provide decision-making tools for growers and crop consultants facing challenging decisions should they experience chemical carryover in their sugarbeet crop.

ANFINRUD, MARK* and DAN BJUR, SESVanderHave USA, 5908 52nd Ave S, Fargo, ND 58104.

Genetic variability creates difficulties in scalping research trials.

Defoliation practices such as scalping and flails are valuable tools in commercial operations of US sugarbeet growers. It's well documented that the top of the beet (crown, petiole, leaves) crown carries more impurities that increase loss to molasses and negatively impact storability. Commercial growers, who generally have a single variety in each field, easily adjust their defoliators and scalpers for the most effective removal of material that would negatively impact their sugar payment while preserving as much of their yield as possible. Research trials are utilized to test many different varieties side by side, which, due to genetics, can vary drastically in crown height and shape. Scalping each variety effec-

tively in yield trials is difficult, as varieties with taller crowns are scalped more aggressively than those with shorter crown heights. Additionally, varieties with narrow crowns have less material removed compared to varieties that produce wider crowns. These differences can be further magnified seed quality differences, as gaps in stands contribute to more extreme crown height variance. Many research trials utilize a subsample system for quality analysis. In this system, 20 to 30 pounds, or about 10-15 roots, are sampled from a research plot. These samples are brought to a tare lab for analysis to determine the quality of the variety. Tare labs handle these samples differently from market to market. Some labs hand trim the roots of petiole material prior to processing that may have been missed by the defoliator/scalper in the field, while other labs simply process the roots as they are. Due to genetic diversity, it's very difficult to evenly scalp each variety in today's research trials. Our studies show losses of over 2 tons/acre due to over scalping varieties with taller or wider crowns. On the contrary, lower profile varieties often get very little scalping which increases the impurity aspect of those varieties and subsequently reduces recoverable sugar per ton. Research teams always try to minimize variables that can impact results. Our findings show that scalping variance can impact a research team's ability to report the most accurate assessment of a variety's genetic potential.

BUBLITZ, DANIEL* and THOMAS WENZEL, Sugarbeet Advancement, Michigan State University, 3775 S Reese Rd, Frankenmuth, MI 48734.

Investigating the impact of clover and radish cover crops on sugarbeet yield in a sugarbeet cyst nematode environment.

Crop rotation is an important aspect of sugarbeet production. In Michigan, wheat is the most common crop to plant the year before sugarbeet, but corn is also popular, followed by soybeans, dry beans, alfalfa, and cucumbers. An important but at times overlooked aspect of crop rotation is the use of cover crops. Incorporating cover crops into a crop rotation offers a number of benefits, such as increasing soil organic matter, improving soil water retention, providing additional nutrients such as nitrogen, reducing soil erosion, improving soil health and microbial activity, and mitigating certain pests. For Michigan sugarbeet growers, cover crops are most often included in the rotation after wheat. Two popular cover crops planted are mixes of clover species, and sugarbeet cyst nematode trap crop species of oilseed radish. Both cover crops offer a unique set of benefits to sugarbeets. The primary benefits of clover are to increase organic matter and provide additional nitrogen, while the trap crop radish can break up soil compaction and decrease the population of sugarbeet cyst nematodes. In the current study, the impacts of clover and radish cover crops on sugarbeet yield the following year were compared, along with a check that had no cover crop planted. This is a multi-year study for which data collection began in 2020 and is ongoing. A different field location has been used every year, each of which has a known history of sugarbeet cyst nematode. The effects of the different treatments were determined using several metrics, including measuring nematode presence at the end of the season, conducting dead beet counts to assess the influence each treatment had on root disease, and recording yield and sugar data. Yield and sugar data were used to calculate the grower's gross income with each treatment. The costs associated with the different cover crop systems were then used to calculate the grower's net income for each. Selecting the appropriate cover crop is an important decision for growers, as it could have an impact on farm revenue and profitability.

BURT, EMMA L.*[1], THOMAS J. PETERS[2], ALEXA L. LYSTAD[2] and DAVID C. METTLER[3], [1]Minn-Dak Farmers Cooperative, 7525 Red River Road, Wahpeton, ND 58075 [2]North Dakota State University and University of Minnesota, NDSU, Dept. 7670, PO Box 6050, Fargo, ND 58108-6050, [3]Southern Minnesota Beet Sugar Cooperative, 83550 County Road 21, Renville, MN 56284.

Our experiences with acifluorfen in sugarbeet.

Acifluorfen for control of glyphosate resistant (GR) waterhemp (*Amaranthus tuberculatus*) in sugarbeet was evaluated in Minnesota and North Dakota in 2019, 2020, and 2021. Acifluorfen at 0.28 kg ha^{-1} plus non-ionic surfactant (NIS) at 0.125% v/v or acifluorfen mixtures with glyphosate (PowerMax) at 0.28 + 1.10 kg ha^{-1} controlled waterhemp escapes up to 10-cm tall once sugarbeet reached the 6-lf stage using ground application methods in a minimum of 140 L ha^{-1} water carrier with nozzles delivering a medium droplet spectrum. Acifluorfen was applied on over 150,000 ha following Environmental Protection Agency (EPA) approval of a Section 18 emergency exemption in 2021 and 2022. In addition to waterhemp size, sugarbeet must be greater than the 6-lf stage at application to avoid detrimental sugarbeet injury and root yield loss. Evaluation of sugarbeet tolerance and waterhemp control is on-going. The likelihood for sugarbeet injury was greater when day-time maximum air temperatures were greater than 29C, when sugarbeet was less than 6-lf stage, and/or when glyphosate was mixed with acifluorfen. Likewise, waterhemp regrowth has been observed when waterhemp size was greater than 10-cm or when sugarbeet partially shielded targeted waterhemp. Field experiments in 2022 a) evaluated sugarbeet tolerance from single or repeat acifluorfen application alone or in mixtures with glyphosate and adjuvants and b) evaluated optimizing carrier volume and spray nozzles to improve waterhemp control. Experiments were planted at seven locations in Minnesota between May 16 and June 3, 2022. Sugarbeet injury, root yield, and percent sucrose from glyphosate (PowerMax3) at 1.05 kg ha^{-1} with NIS and ammonium sulfate was applied at the 2- and 6-lf sugarbeet stage and was compared to sugarbeet injury, root yield, and percent sucrose from a single acifluorfen application at 0.28 kg ha^{-1} with NIS at 0.25% v/v or crop oil concentrate at 1.17 L ha^{-1} at the 6-lf stage, acifluorfen at 0.21 kg ha^{-1} with NIS at 0.125% v/v at the 6-lf stage followed by a repeat acifluorfen application and NIS seven days later, and acifluorfen applied in mixtures with glyphosate with AMS or with AMS and NIS at the 6-lf stage. Repeat acifluorfen applications or acifluorfen mixed with glyphosate increased sugarbeet necrosis or growth reduction injury as compared to glyphosate. However, acifluorfen alone did not reduce root yield as compared to the glyphosate control treatment. Acifluorfen mixtures with glyphosate reduced or tended to reduce root yield and recoverable sucrose ha^{-1} but did not reduce percent sucrose as compared to glyphosate or acifluorfen alone. Carrier volume or spray nozzle did not affect sugarbeet injury, although acifluorfen applied through flat fan nozzles tended to increase sugarbeet injury as compared to other nozzles evaluated. Waterhemp control was best when acifluorfen was applied at 187 L

GILDER, JEFF*, The Andersons, Inc., 1947 Briarfield Blvd, Maumee, OH 43537.

Effect of potassium acetate on sugar beets late season plant health.

Potassium is an essential nutrient in sugar beet production as it plays a major role in the translocation of sugars and helps prevent energy loss in the plant by reducing respiration. A late season deficiency can reduce the overall quality and yields of sugar beets. The Andersons conducted trials in Michigan and the central Red River Valley with the goal of improving late season plant health through the application of potassium acetate. Results from the research trials demonstrate that a foliar application of potassium in an acetate form increases RWST. Application timing was 12-leaf followed by 30 days prior to harvest. In the trial, The Andersons Korrect® and Korrect Plus were applied. Both products are highly available mild forms of potassium which are particularly well suited for foliar feeding for all types of crops and may be soil applied. With a pH of 7.3 and salt index of 26, there is little danger of burn to plant leaves and foliage. Korrect and Korrect Plus are clear, true solutions with no particulate matter, which mix readily in water. The natural organic carrier enhances its receptivity by plants. The products contain no chlorides or ammonium (NH_4) forms of nitrogen. Potassium acetate is the most available form of potassium to use in a foliar application. Translocating sugars in the plant helps lead to increased yield and RWST.

HILSCHER, ELKE*, HEIKO NARTEN and TORBEN ERICHSEN, KWS SAAT SE & Co. KGaA, Grimsehlstraße 31, Einbeck, 37574 Germany.

How to produce a representative sugarbeet sample.

Comparison to the current analytical method is one of the main concerns during implementation of KWS BEETROMETER in sugar factories. In the development of a new method with potential advantages (for example cost reduction, improvement in efficiency, easier procedures and maintenance) it is important to proof that the standard, brei-based lab method can be replaced by NIRS without compromising analytical quality. Differences between methods and processes are due to analytical deviations and therefore an understanding of the reasons for deviations are essential. The heterogeneity in sugarbeets has the largest impact on the estimation of the beet quality. Sampling is a critical component in any analytical procedure, and one must always ensure representativity. Sampling errors typically can reach up to 100 times the specific analytical error. "Incorrect sampling" always results in non-representative samples and thus to an uncontrollable bias even though all subsequent steps have been performed in a correct way [(1)]. Results are presented that show which influencing factors must be considered to obtain a representative beet sample for quality analysis. The sample preparation (brei production) is crucial for reliability and reproducibility of sugarbeet quality values. Results of sample preparation experiments are discussed regarding their impact on analytical bias and reason for deviations in analytical results. With the KWS BEETROMETER a standardized solution is available, that allows real-time quality assessment of sugarbeets. As the whole sugarbeet sample is crushed in a patented chopper and subsequently measured a representative sampling is ensured. During measurement with a diode-array NIR spectrometer records a single spectrum every 40 ms thus capturing the whole heterogeneity with approximately 10 spectra per kg. Based on those spectra, a standardized and accurate sample average result can be delivered in real-time.

[1]J.B. Holm-Nielsen et al. / Chemometrics and Intelligent Laboratory Systems 83 (2006) 114–126

NARTEN, HEIKO*, ERICHSEN, TORBEN and HILSCHER, ELKE, KWS SAAT SE & Co. KGaA, Grimsehlstraße 31, Einbeck, 37574 Germany.

Achieving the same prediction quality on different NIR spectrometers.

NIR measurements are an established method for chemical analysis in many industries like the pharmaceutical and agricultural sector. Due to its relative ease of use and higher grade of automation NIR systems are often used in circumstances with decentral and/or multiple lines using more than one spectrometer. Whenever the same quality parameters of a product come from multiple, different instruments precautions need to be taken to minimize deviations between those. This true for any system whether it is NIR-based or relies on wet-chemistry methods like polarization, etc. The KWS BEETROMETER is a NIR-based system that provides real-time quality parameters for sugarbeets and a direct alternative to the classical wet-chemistry based on brei. While wet-chemistry instruments are calibrated with standard samples, other ways must be used with the KWS BEETROMETER as it measures directly on freshly chopped beets – thus perishable samples. Based on more than fifteen years of experience using NIR in sugarbeet breeding, many influencing factors, for example detector and light source characteristics differing for each spectrometer, have been identified. To minimize deviations between systems protocols were developed to ensure a fair payment for both growers and sugar factories based on exact and comparable quality parameters independent of line and location. The poster shows the many precautions and measures involved in ensuring that the KWS BEETROMETER delivers reliable results on different lines, locations and factories to each customer. This involves factors like sample preparation, validation samples as well as calibration transfer and more.

SPRAGUE, CHRISTY L.*, GARY E. POWELL and BRIAN J. STILES II, Michigan State University, 1066 Bogue Street, East Lansing, MI 48824.

Sugarbeet tolerance to acifluorfen (Ultra Blazer) in Michigan.

The lack of postemergence herbicide options to control the glyphosate-resistant *Amaranthus spp.*, Palmer amaranth and waterhemp, is a concern for sugarbeet growers. In the past, sugarbeet has shown some tolerance to the Group 14 herbicide, acifluorfen (*Ultra Blazer*). In soybean, acifluorfen's strengths include control of *Amaranthus spp.*, eastern black nightshade, and common ragweed. Field research (2018-2022) on sugarbeet tolerance to acifluorfen was conducted to justify and support Section 18 labeling of Ultra Blazer in Michigan sugarbeet. Early studies examined acifluorfen combinations with glyphosate (0.84 kg ae ha^{-1}) at rates ranging from 0.14 to 0.43 kg ai ha^{-1} to sugarbeet at the 2-, 6-, and 12-leaf stages. This research showed that acifluorfen should only be applied after sugarbeet was at the 6-leaf stage. We have continued to examine the effects from multiple applications of acifluorfen to 6- and 12-

leaf sugarbeet and combinations with the tank-mixture partners of clopyralid, ethofumesate, s-metolachlor, dimethenamid-P, or acetochlor at 6-leaf sugarbeet. In many cases, the addition of tank-mix partners has increased sugarbeet injury and in the case of acifluorfen combination with ethofumesate sugarbeet yield loss was also observed. When acetochlor was tank-mixed with acifluorfen sugarbeet injury was less. While sugarbeet injury is a concern with acifluorfen, applications to sugarbeet at the 6-leaf stage or greater could be an option for postemergence glyphosate-resistant waterhemp control.

STEINKE, KURT*, LACIE THOMAS, DANIEL BUBLITZ, and JAIME WILLBUR. Dept. of Plant, Soil, and Microbial Sciences, Michigan State University, East Lansing, MI 48824.

Utilizing boron to improve sugarbeet yield, quality, and resistance to *Cercospora beticola*.

Cercospora beticola, the causal pathogen of Cercospora leaf spot (CLS), is one of the more severe foliar pathogens capable of causing damage to sugarbeet. When not managed appropriately, CLS can reduce sugarbeet yield 40%. Although plant defoliation caused by the disease directly impacts root size and sugar quality, other factors including leaf regrowth and impurities within the root affect plant health and crop quality. Management strategies including boron-containing compounds have shown to have fungistatic properties with potential to reduce disease severity in the field. Field studies were established to investigate the effects of foliar applied boron (B) on sugarbeet plant health and CLS disease severity. Treatments included a standard fungicide program, three foliar boron treatments (0.11, 0.28, or 0.56 kg sodium tetraborate ha^{-1}) applied at 10-14 day intervals without a standard fungicide program, three foliar boron treatments (0.11, 0.28, or 0.56 kg sodium tetraborate ha^{-1}) applied at 10-14 day intervals in conjunction with a standard fungicide program, and a nontreated check for a total of eight treatments. Application of foliar B did not reduce CLS in field environments across site years. *In vitro* analysis of *C. beticola* response to B demonstrated lower EC_{50} values with sodium tetraborate than boric acid. Reduced control options, increased CLS resistance, lack of response to previous B applications, and varietal improvements may enhance the need for further evaluation of alternative controls. In-field evaluation of various B timing, increased B concentration, and addition of B-containing compounds may contribute to future CLS control.

STEVENS, BART*, JAY JABRO and BILL IVERSEN, USDA-ARS, 1500 N. Central Ave., Sidney, MT.

Crop residue accumulation and nutrient stratification near the soil surface in tilled and direct-seeded sugarbeet cropping systems.

Rising tillage costs have increased interest in reduced tillage among sugarbeet (*Beta vulgaris* L.) growers. No-tillage (NT) sugarbeet planting offers cost and soil conservation benefits but also nutrient management challenges. Growers transitioning from tillage to NT who do not have the equipment to sub-surface band fertilizer often resort to broadcast applications without incorporation. This will likely lead to nutrients being concentrated near the soil surface but sugarbeet harvest causes some soil disturbance and may ameliorate stratification. Similarly, NT allows crop residue to accumulate on the soil surface providing protection against soil erosion and a reservoir of slow-release nutrients. A field study was conducted in North Dakota, USA from 2013 to 2019 to evaluate the effect of NT on nutrient stratification in 2-year (barley [*Hordeum vulgare* L.]/sugarbeet) and 4-yr (corn [*Zea mays* L.]/soybean [*Glycine max* L.]/barley/sugarbeet) rotations under sprinkler irrigation. A corn-soybean rotation was included to provide a comparison with a rotation that doesn't include sugarbeet. Tillage (30 cm deep) was performed on subplots in each of the rotations as a control. Tillage treatments were applied consistently from year to year to all rotation crops on a given plot. Soil samples were collected following barley (fall 2019) and corn (spring 2020). Soil cores were divided into depths of 0-5, 5-10, 10-15 and 15-20 cm and analyzed for extractable P, exchangeable K and organic matter. Residue samples were collected after the 8-year study was completed and residue biomass, total N and total C were determined. Stratification was greater with NT than with tillage in all rotations; however, the concentration of nutrients in the top 5 cm of soil was somewhat less in NT rotations that included sugarbeet than in the corn-soybean rotation. Stratification was greater for the largely immobile P and OM than for the somewhat mobile K. Greatest accumulation of OM in the top 5 cm of soil occurred in the NT corn-soybean plots, likely as a result of the large amount of corn residue returned to the soil every other year in this system.

TARKALSON, DAVID*[1], DAVEY OLSEN[2] and DAVE BJORNEBERG[1], [1]USDA-ARS, 3793 N 3600 E Kimberly, ID 83341-5076, [2]Amalgamated Sugar Company, 1951 S. Saturn Way, Suite 100, Boise, ID 83709.

Effect of nitrogen supply by soil depth on sugarbeet production and quality.

Determining the effect of N supply by soil depth on sugarbeet production is important to continue fine-tuning management practices. To accomplish this objective, a greenhouse column study was conducted at USDA-ARS in Kimberly, ID. The study was conducted using 30 1 m x 0.3 m columns filled with 0.9 m of soil. The treatments consisted of adding N fertilizer at a rate of 132 kg N ha^{-1} differently to three 0.3 m soil depths. Although all treatments had a total N supply of 222 kg N ha^{-1} in the 0.9 m soil depth, the distribution of the N in the soil profile affected the yields and N uptake. Sugarbeet root mass, root sucrose mass, leaf mass, root N mass, and leaf N mass were higher for treatments where N fertilizer was added to depths 1 (0-0.3 m) and 2 (0.3-0.6 m) compared to when N fertilizer was added to depth 3 (0.6-0.9 m). The sugarbeets were not able to utilize N from depth 3 as efficiently as from depth 1 and depth 2. There were no treatment effects on sugarbeet quality factors. The findings of this study highlight the need to question the value of a depth 3 soil sample for determining N fertilizer requirements.

TARKALSON, DAVID*[1], DAVEY OLSEN[2], DAVE BJORNEBERG[1], GREG DEAN[2] and CLARKE ALDER[2], [1]USDA-ARS, 3793 N 3600 E, Kimberly, ID 83341-5076, [2]Amalgamated Sugar Company, 1951 S. Saturn Way, Suite 100, Boise, ID 83709.

Static range nitrogen management in northwest U.S. sugarbeet production.

Nitrogen (N) management is important in sugarbeet production due to effects on yield and quality. Evaluation of recent and past research data from the Northwest U.S. production area shows that using a static range N management (SRNM) approach better matches sugarbeet N needs than currently used the YGNM approach. The SRNM approach uses a range of fixed N supply independent of yield goal. The term "range" is used due to slight variations in required N supply due to soil types, climate, and other fixed factors. This paper will provide evidence for a shift from a YGNM approach to a SRNM approach. □

ORAL PRESENTATIONS

BOLTON, MELVIN D.*[1], LORENA I. RANGEL[1], MARI. B. NATWICK[2], GARY A. SECOR[2] and VIVIANA RIVERA-VARAS[2], [1]USDA-ARS, 1616 Albrecht Blvd N, Fargo, ND 58102, [2]North Dakota State University, Department of Plant Pathology, NDSU Dept 7660 PO Box 6050 Fargo, ND 58108.

CbCyp51-mediated DMI resistance is modulated by codon bias.

Cercospora leaf spot (CLS) is the most damaging foliar disease of sugar beet globally. To combat CLS, multifaceted efforts are widely employed, including breeding for resistance, cultural practices, and the application of fungicides. However, populations of *Cercospora beticola* have become resistant to most fungicides used for CLS management, including those in the sterol demethylation inhibitor (DMI) class of fungicides. In this study, we sampled nearly 600 isolates of *Cercospora beticola* from MN and ND during the 2021 sugar beet growing season. For each isolate, EC_{50} values were determined for DMIs tetraconazole (Eminent), prothioconazole (Proline), difenoconazole (Inspire), and mefentrifluconazole (Revysol). Using the *CYP51* gene sequence for each isolate, we determined that the synonymous E170 mutation and the synonymous/nonsynonymous L144(F) can be used to predict resistance to these four DMIs. The prevalence and accuracy of the six mutation combinations were calculated and specific combinations can predict resistance with greater than 90% accuracy. Interestingly, one prevalent mutation combination resulted identified cross-resistance to difenoconazole and mefentrifluconazole, but sensitivity to tetraconazole and prothioconazole. This data reveals the importance of codon bias in fungicide resistance and is the first demonstration of the use of synonymous mutations to predict cross-resistance.

CAMELO-GARCÍA, VIVIANA*, WINTERMANTEL, WILLIAM and MARTIN, FRANK, USDA-ARS, 1636 E Alisal St., Salinas, CA 93905.

Study the genetic variation of *Polymyxa betae* using targeted amplicon sequencing.

Polymyxa betae, a plasmodiophorid protist, is the vector of the soilborne virus beet necrotic yellow vein virus (BNYVV), the cause of rhizomania disease and the most economically important virus of sugar beet. Molecular studies of these organisms are difficult due to their biotrophic nature; thus, there is still much to be learned about them. In this study, we attempted to develop molecular methods to analyze the genetic variation in seventeen specific regions of the *P. betae* genome using the Illumina MiSeq platform with a two-step PCR approach. The target-specific regions, which contain single nucleotide polymorphisms and insertions/deletions, were identified previously through comparative analysis of genome assemblies from ten isolates recovered from soil samples collected from sugar beet growing areas across the U.S.A. The variation in those target-specific regions was previously validated using a set of twelve genotyping markers. In this preliminary assay, we used *P. betae* DNA extracted from sugar beet roots and soil, to compare the total sequencing output using different DNA sources. Read mapping and variant calling allowed the identification of the SNP and INDEL observed previously using the genotyping markers. The total reads count for each of the seventeen target-specific amplicons for the soil samples were slightly lower than for sugar beet roots, but had no impact on the type or frequency of variants detected as they were similar in both types of samples. The pilot study demonstrated the capability of using the newly developed genotyping markers in MiSeq analysis to study the genetic variation of *P. betae*. Currently, we are evaluating the technique on soil samples collected from a sugar beet production area in Imperial Valley, CA.

CHU, CHENGGEN*[2], ZHENG, YAOJIE*[1], TEHSEEN, MUHAMMAD MASSUB [1], and LI, XUEHUI[1]., [1]North Dakota State University, 1420 Bolley Dr, Fargo, ND 58105, [2]USDA-ARS, 1616 Albrecht Blvd N, Fargo, ND 58102.

Development of DNA STARP marker platform for flexible SNP genotyping in sugarbeet.

Whole-genome sequencing enables high-throughput detection of SNP (single nucleotide polymorphism) covering the whole genome among plant individuals simply by comparing DNA sequences, which greatly accelerates the identification of associations between specific SNP markers with the target traits through genome-wide association studies (GWAS). Tightly associated SNPs thus can be efficiently used for marker-assisted selection (MAS) during trait integration or locating candidate genes for gene isolation. This requires developing a SNP genotyping system to flexibly detect specific SNP alleles at a few associated loci. Several PCR-based platforms for individual SNP genotyping were developed but most of them used only the fluorescence signal to differentiate SNP alleles. The allele discrimination through fluorescence signal could be less accurate if the strong background signal noise was presented. A newly developed STARP (semi-thermal asymmetric reverse PCR) marker system increased the sensitivity of discriminating SNP alleles by introducing more polymorphic nucleotides in its forward primers, which thus increases the specificity of PCR products from corresponding alleles and the polymorphisms can be detected using either fluorescence signal or the traditional gel electrophoresis system that relies on DNA fragment size difference. The STARP marker system will be more preferrable for flexible SNP genotyping in sugarbeet since the widely spread repetitive sequences and a high level of heterozygosity in the sugarbeet genome produced strong background signal noise. In this study, genotype-by-sequencing (GBS) was used to detect SNPs among sugarbeet lines, and the selected SNPs were further verified through sequencing PCR products amplified from the corresponding regions. STARP primers were designed according to sequences harboring SNPs, and the resulting PCR products clearly discriminated SNP alleles between sugarbeet lines and their F_1 hybrids using both methods of fluorescence and gel electrophoresis. This study proved the efficacy of the STARP markers for sugarbeet SNP allele discrimination and developed a flexible SNP genotyping platform in sugarbeet.

CURRAN, MICAH T.*, JOHN J. WEILAND and MELVIN D. BOLTON, USDA-ARS, 1616 Albrecht Blvd. N Fargo, ND 58102-2765.

Protection against Beet mosaic virus via topical application of dsRNA.

Beet mosaic virus (BtMV), a *Potyvirus* found worldwide capable of causing reduction in sugar yield, was studied because of its frequent and synergistic interactions with the virus complex that comprises Virus Yellows, an important disease of sugar beet. Facing potential regulations for current management strategies for this disease, there is an increasing need for environmentally friendly new age pesticides to manage Virus Yellows and other important insect-vectored viruses of sugar beet. Here, double stranded RNA (dsRNA) was studied as an alternative control strategy by employing RNA interference (RNAi) to inhibit virus replication in sugar beet and *Chenopodium quinoa*. DsRNA targeting the BtMV 5' UTR, 3' UTR, and virus coat protein were constructed and tested as targets to disrupt viral replication and infection. In all experimental replicates in *C. quinoa* as a local lesion host, negative and nonspecific dsRNA control treatments resulted in severe infection indicated by a high number of lesions. In contrast, inoculated *C. quinoa* treated with a low dose of dsRNA (100 pmol) specific to the BtMV 5' UTR, 3' UTR, and coat protein demonstrated a reduction in local lesions up to 98.8%, 95.7%, and 94.1%, respectively. Lower quantities of dsRNA were tested with a slight decrease in control indicating that control is dose dependent. Interestingly, BtMV was found to translocate systemically in *C. quinoa*, a characteristic previously unreported for this virus. Systemic infection in otherwise protected *C. quinoa* may indicate that mobile RNAi is required for full plant protection in this pathosystem. Time course and dose dsRNA studies will be discussed. Preliminary results suggest that dsRNA has potential as a disease management strategy for Virus Yellows.

DORN, KEVIN*, USDA-ARS, Crops Research Lab, 1701 Centre Ave., Ft. Collins, CO 80526.

Rapid trait mapping and characterization with purpose-built genome assemblies.

Genomics has revolutionized plant breeding in the past decade and continues to rapidly shift the definition of state-of-the-art of the field. The adoption of genomics-assisted methods in both public and private sugar beet breeding efforts has been substantially bolstered by the release and continued updates of both the RefBeet and EL10 sugar beet genome assemblies. These two foundational reference genomes for sugar beet have enabled wide-scale adoption of high throughput genotyping and whole genome sequencing to study population genetics, resistance trait discovery and characterization, and domestication. The rapid advancement of sequencing technology, coupled with plummeting sequencing costs, have democratized plant de novo genome assembly. In this talk, I present examples of how three new beet genome assemblies that were built to mine specific traits of interest from specific genotypic backgrounds to specifically avoid reference genome bias. These examples each represent a new, tractable and affordable approach for the rapid mapping of traits of interest from any line of sugar beet or crop wild relative. Each of the three genome assemblies, two from Beta vulgaris and one from Patellifolia procumbens, are chromosome scale and fully phased. All three assemblies are the most complete and contiguous assemblies to date, and are enabling the beginnings of pan-genome analyses in beet, as well as shedding light on the genome biology and evolutionary history of the Betoideae subfamily.

EINI OMID* and VARRELMANN MARK, Institute of Sugar Beet Research, Department of Phytopathology; Institute of Sugar Beet Research; 37079 Göttingen, Germany.

Response of sugar beet cultivars to Beet curly top virus infection: virus accumulation and transcriptome assay.

Curly top disease caused by geminiviruses including *Beet curly top virus* (BCTV) and *Beet curly top Iran virus* (BCTIV) is a limiting factor for sugar beet production. In recent years, climate change has increased the spread of virus vectors and disease damage in infected regions. The most economical control of BCTV in sugar beet would be through resistant cultivars though, most commercial cultivars possess only low to moderate resistance. A doubled haploid Line KDH13 showed resistance to BCTV and produced no symptoms. However, the response of Line KDH13 to BCTIV was not studied before. Here we tested the response of Line KDH13 to the BCTIV infection and compared with BCTV infection by means of inoculation with infectious clones mobilized by agrobacteria. Real-time PCR showed that both viruses replicated in the locally agroinfiltrated cotyledons. The viral DNA accumulation was lower for BCTIV compared to the BCTV infection. Systemically infected resistant plants with BCTV showed mild enation without leaf curling after 30 days. In contrast, leaf curling appeared after 12 days in the susceptible line. Further, BCTIV was not detected in resistant plants. Transcriptome analysis for the BCTV infected plants showed only 43 genes that were deregulated in line KDH13 compared to the regulation of a large number of genes (1139 gene) in the susceptible line. This work demonstrates the response of sugar beet plants to BCTV / BCTIV infection at both local and systemic infection and highlight the metabolic pathways/defense-related genes for their contribution towards BCTV resistance.

FLOBINUS, ALYSSA*[1], JOHN J. WEILAND[1], CARLA M. REIS VARANDA[2], PATRICK MATERATSKI[2], GARY A. SECOR[3] and MELVIN D. BOLTON[1], [1]USDA-ARS, 1616 Albrecht Blvd N, Fargo, ND 58102, [2]University of Évora, Laboratory of Plant Virology, Évora, 7006-554, Portugal, [3]North Dakota State University, Department of Plant Pathology, NDSU Dept 7660 PO Box 6050 Fargo, ND 58108.

Characterization of *Beta vulgaris* Alphanecrovirus-1, a putative Alphanecrovirus identified in sugar beet.

Beta vulgaris Alphanecrovirus-1 (BvANV-1) was recently identified in our laboratory through RNA sequencing analysis of infected sugar beet roots. To determine the impact of BvANV-1 on sugar beet and other locally-grown rotational crops and weeds and to evaluate viral transmission by the *Olpidium* sp. vector, infectious clones were developed. Two variants of BvANV-1 were obtained; both clones share a similar sequence with the exception of three single nucleotide polymorphisms (one amino acid substitution and two silent mutations represented in the BvANV-1-7 clone) compared to the putative wild-type sequence found in BvANV-1-10 clone. Infection of *Nicotiana benthamiana* with BvANV-1-10 led to strong, systemic necrosis on all leaves and stunting of the plant, while plants inoculated with BvANV-1-7 exhibited infrequent necrotic lesions on inoculated leaves. To determine which mutation(s) were involved with the decreased virulence on BvANV-1-7-inoculated *N. benthamiana*, different combinations of point mutations were introduced reciprocally from BvANV-1-10 into BvANV-17 clone and tested using mechanical inoculation of transcript RNA. The impact of the mutation on aggressiveness will be presented along with other data regarding characterization of this putative pathogen of sugar beet including host range and viral transmission by the vector.

HANSON, LINDA E.*[1] and DOUGLAS H. MINIER[2], [1]USDA-ARS, 612 Wilson Road, 37 Plant Biology Lab, East Lansing, MI 48824, [2]Michigan State University, 612 Wilson Road, 37 Plant Biology Lab, East Lansing, MI 48824.

Histopathology of *Rhizoctonia solani* AG 2-2 infection of sugar beet.

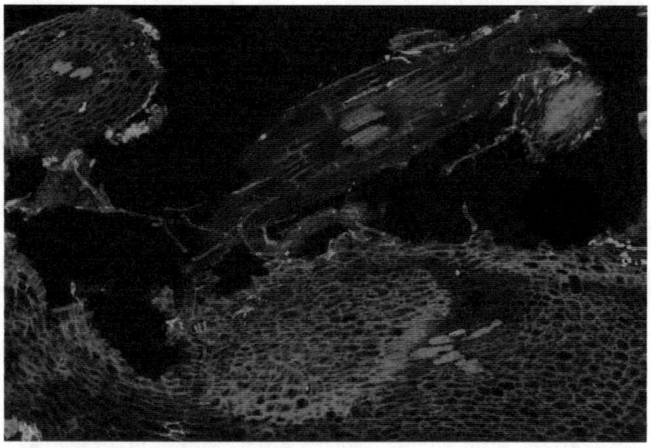

The infection process for *Rhizoctonia solani* AG 2-2 was investigated by Ruppel in 1973, with two early resistant and two susceptible germplasm. Since that time, there have been advances in microscopy and detection methods. The aim of this work was to use advanced microscopy and detection methods to reexamine infection processes in sugar beet, evaluate the response of a modern resistant (highly resistant) and a susceptible germplasm, and to compare those responses with observations made in the 1970s. As in Ruppel's work, infection on the root tissue occurred via infection cushions, but the presence of infection cushions was not always evident and direct penetration without forming an infection cushion is likely. Fungal hyphae that entered through the root surface were largely restricted to the outermost cortical layer in both the resistant and sensitive varieties. The outermost cambial ring appeared to act as a barrier to fungal progression, with hyphae invading only through breaks in the cambial ring where non-vascular tissue was present. However, within the root groove, fungal hyphae were observed that had entered, evidently by direct penetration, the tissues where feeder roots emerged. This infection strategy allowed deeper penetration of the fungus into the root, bypassing the potential barrier presented by the outermost cambium. As reported by Ruppel, symptoms were found in advance of the fungal hyphae. Observations of necrosis were consistent with a breakdown of plant tissue by fungal polygalacturonases. Increased affinity for propidium iodide in the area surrounding the invading hyphae indicated possible de-esterification of the cell walls, which may be a precursor to degradation by pectinases. Results did not support a lysis of fungal tissue as hypothesized by Ruppel. Instead, plant physical and chemical barriers were evident.

KHAN, MOHAMED F. R.*[1,2] and PETER HAKK[1]. [1]Plant Pathology Department, North Dakota State University and [2]University of Minnesota, Fargo, ND 58108-6050.

Effective and economic management of *C. beticola* using timely fungicide applications and host resistance.

Cercospora beticola causes Cercospora leaf spot (CLS), the most economically destructive disease of sugar beet in the North Central USA. From 1999 through 2015, most growers in Minnesota and North Dakota used quinone outside inhibitor (QoI), demethylation inhibitor (DMI), triphenyltin hydroxide (TPTH), and to a lesser extent thiophanate methyl in an alternation program to successfully manage sensitive populations of *C. beticola*. Over time, the pathogen developed resistance to most of the fungicides used for its control. In 2016, growers in the USA lost over $200 million because of QoI resistance, reduced sensitivity to DMI fungicides and a CLS epidemic. There is no individual fungicide that currently provides effective control of *C. beticola* and no new chemistry identified for CLS control in the near future. Sugar beet seed companies have been working at developing varieties with improved tolerance to *C. beticola*. Current field trials in the USA indicate that new cultivars (CR+) have better resistance to *C. beticola* compared to the currently approved non-CR+ cultivars. Field trials using currently approved moderately tolerant varieties (non-CR) and new improved CR+ varieties were conducted in 2019 through 2022. Fungicide applications were done on a calendar basis, and on an only when needed based on the presence of symptoms and favorable environmental conditions. In 2019, under environmental conditions favorable for fungicide applications, the newer improved CR+ varieties provided higher recoverable sucrose than one non-CR+ variety and similar yields to a second non-CR+ variety. In 2020, under conditions favorable for CLS, the new CR varieties produced significantly higher recoverable sucrose compared to the non-CR varieties. One CR-variety was highly resistant to *C. beticola* and its yield of recoverable sucrose in the absence of fungicides was not significantly different from treatments that received multiple fungicide applications. In a dry and warm 2021 season, fungicides provided protection for the non-CR varieties, but were not necessary the CR+ varieties. In 2022, conditions were not favorable for development of *C. beticola* until late August. One or two timely fungicide applications based on the presence of symptoms and thresholds resulted in low disease severity and recoverable sucrose similar to treatments with five applications that started before or at row closure. The availability and use of the newer CR+ varieties will improve the economic viability of the sugar beet industry. However, fungicides will have to be used judiciously to prolong the usefulness of the CR+ varieties.

KUMAR, AVNEESH*[1], ANN FENWICK[2], AMY NIELSON[2] and KEVIN DORN[2], [1]Department of Agricultural Biology, Colorado State University, 1701 Centre Ave, Fort Collins, CO 80526 [2]Soil Management and Sugarbeet Research, USDA-ARS, 1701 Centre Ave, Fort Collins, CO 80526.

BeetBase: A sugar beet toolbox for genetics, genomics, and germplasm research.

BeetBase, the USDA-ARS web-based platform sugar beet database, provides germplasm characterization information including genotyping, phenotyping and whole genome sequencing datasets, genome assemblies, and annotations as well as providing powerful computational tools to efficiently utilize this genomic information. It is aimed to bring the usefulness of publicly available datasets by parsing data through established workflows and providing the dataset in an easy-to-use format. In version 2 of BeetBase, we add functionality and datasets to enable users to mine the genetic diversity of sugar beet and crop wild relatives. New tools include whole genome sequencing datasets, variant analysis data and tools, haplotype mining, synteny analysis between multiple reference genomes, and germplasm data mining. We envision these new BeetBase functionalities providing a platform for open data and data reuse, improving the visibility of ongoing research efforts.

KURODA, YOSUKE*[1], MATSUHIRA, HIROAKI[1], NARIHIRO, TSUBASA[1], SANO, MASAKAZU[1], KITAZAKI, KAZUYOSHI[2] and KUBO, TOMOHIKO[2], [1]Hokkaido Agricultural Research Center, National Agriculture and Food Research Organization, 9-4 Shinseiminami, Memuro, Kasai, Hokkaido 082-0081, Japan, [2]Laboratory of Genetic Engineering, Research Faculty of Agriculture, Hokkaido University, Sapporo 060-8589, Japan.

Resistant breeding for Virus yellows in Japan.

Virus yellows (VY) is an important disease of sugar beet, caused by some virus species transmitted by the polyphagous aphid *Myzus persicae* and others, which causes yellowing symptoms on the outer large leaves and reduced sugar yield. The VY prevalent in Japan was previously thought to be Beet Western Yellowing Virus (BWYV), which has been identified in the United States, but was recently named Beet Leaf Yellowing Virus (BLYV), a new species in the same genus (Polerovirus) as BWYV. In this study, we report the following three findings for the aim of efficient resistant breeding for BLYV. (1) Through four years of preliminary research, an assay was developed to estimate leaf yellowing caused by BLYV using an index (0: no symptoms – 3: overall yellowing) or the yellowing area ratio (%) calculated from digital camera images. (2) Using this assay, a two-year evaluation of 100 Japanese lines including USDA VY-resistant germplasm revealed that only the germplasm showed strong resistance to BLYV, suggesting that the mechanism of resistance to BWYV and BLYV may be common. (3) QTL analysis was performed from phenotype and genotype data of 144 $F_{2:3}$ progenies developed by crossing susceptible and resistant parents for two years, and several common QTLs associated with the resistance were detected. DNA markers were developed to select for BLYV resistance based on the information. This research was supported by the research program on development of innovative technology grants (Grant Number: JPJ007097) from the Project of the Bio-oriented Technology Research Advancement Institution (BRAIN).

LIEBE, SEBSTIAN*[1], EDGAR MAISS[2] and MARK VARRELMANN[1], [1]Institute for Sugar Beet Research, Department Phytopathology, Holtenser Landstr. 77, D-37077 Göttingen and [2]Leibniz University Hannover, Institute of Horticultural Production Systems, Dept. Phytomedicine, Herrenhäuser Str.2, 30419 Hannover.

Adaption of BNYVV towards *Rz1* resistance in sugar beet

The beet necrotic yellow vein virus (BNYVV) causes the rhizomania disease in sugar beet, which is controlled since more than two decades by varieties harboring the Rz1 resistance gene. The development of resistance-breaking strains has been favoured by a high selection pressure on the virus population. Resistance-breaking is associated with mutations at amino acid positions 67-70 (tetrad) in the pathogenicity factor P25 and the presence of an additional RNA component (RNA5). However, natural BNYVV populations are highly diverse with more than 25 known tetrad variants and different RNA5 species making investigations on the resistance-breaking mechanism rather difficult. Therefore, we refined our previously developed reverse genetic system for BNYVV (A type) to study the *Rz1* resistance-breaking mechanism by direct agroinoculation of sugar beet seedlings. Our system allowed a clear discrimination between susceptible and resistant genotypes. A subsequent screen for resistance-breaking revealed that multiple tetrad variants allow BNYVV to overcome *Rz1*. Furthermore, certain mutations severely impaired the viral pathogenicity in the susceptible genotype suggesting a crucial function of the tetrad in P25. Finally, the supplementation of an additional RNA5 species, either from the J or P type group, allowed virus accumulation in the *Rz1* genotype independent of the P25 tetrad. However, this effect was impaired in a genotype carrying *Rz2*, which is another resistance gene against BNYVV. Our infection system enables a rapid identification resistance-breaking mutations and highlights the plasticity of the BNYVV genome allowing adaption towards *Rz1*.

LEIN, JENS C.*[1], BEYER, WERNER[1], WEISSLEDER, KNUTH[1] and REKOSKE, MARGARET M.[2], [1]KWS SAAT SE & Co. KGaA, Grimsehlstraße 31, 37574 Einbeck, Germany, [2]KWS Seeds, LLC, 1325 Valley View Rd, 55379 Shakopee, U.S.A.

How to model sugar beet to face next decade's challenges in sugar production – a perspective from breeding.

Since the development of early sugar beet varieties as the main sucrose producing crop of temperate climate zones in the mid of the 19th century, a continuously breeding driven adaptation to significantly changing agriculture practices as well as to newly arising pests and diseases was key for the success of the crop until today. Milestones in these developments were for example: the shift from multigerm to monogerm varieties, innovative rhizomania and nematode tolerant hybrids or Glyphosate- and Sulfonylurea herbicide tolerance systems in North America or Europe, respectively. In addition, continuously increasing sugar yields combined with elevated sugar contents ensured the competitiveness of sugar beet in the various cropping regions and made sugar beet to a stable and highly profitable crop for the whole value chain. Looking to the next decades new challenges appear on the horizon and are threatening the stability and profitability of beet-based sugar production. Increasing extreme weather events caused by climatic changes, restrictive frame-conditions for farming or missing next generation plant protection products can destabilize the production systems with different significance regarding the worldwide production areas. To encounter these threats, the identification of new genetic solutions is key to keep sugar beet as an important factor in the respective crop rotations. New and innovative test-systems for biotic and abiotic stresses need to be developed as a means to fulfill these important goals. In addition, molecular and environmental information will expand predictive breeding approaches for known and newly identified genetic variation and with this opening the door for sophisticated crop modelling. Consequently, KWS has invested in the past years significantly in the establishment of technologies, test-systems and breeding approaches with the goal to speed-up breeding and extend biotic and abiotic stress resistance. In this way we will model sugar beet to a new level of robustness and keep it as a highly profitable crop in the up-coming challenging decades.

LIEN, AUSTIN* and ASHOK CHANDA, Dept. of Plant Pathology & Northwest Research and Outreach Center, University of Minnesota, U of M NWROC, 2900 University Ave., Crookston, MN 56716.

Tank-mix partners selectively enhance DMI fungicides in

controlling Cercospora leaf spot and impact the genetic diversity of a *Cercospora beticola* population.

Cercospora leaf spot (CLS) caused by *Cercospora beticola* is considered the most destructive foliar disease of sugarbeet worldwide. With limited availability of commercial sugarbeet varieties having strong tolerance to CLS, the use of fungicides remain critical for managing CLS and reduce the extent of economic losses. Demethylation inhibitor (DMI) fungicides are currently the most important group of fungicides used to manage CLS in sugar beets. Additionally, combining DMI fungicides with a broad-spectrum tank-mix partner is highly recommended to delay the development of fungicide resistance. A two-year field study (2020 and 2021, artificially inoculated) was conducted to evaluate the efficacy of DMI fungicides with and without tank-mix partners for managing CLS. A 5 x 7 factorial split-plot design with four replications was used. Treatments included four DMI fungicides (prothioconazole, tetraconazole, difenoconazole + propiconazole, and mefentrifluconazole) and six tank-mix partners (mancozeb, copper oxychloride + copper hydroxide, sodium bicarbonate, sulfur, potassium phosphite, and *Bacillus subtilis*). CLS disease severity was quantified using APS Assess 2.0 software. In both years, significant differences were observed for the tank-mix partners with mancozeb, copper, potassium phosphite, and sulfur resulting in increased recoverable sucrose compared to DMI fungicides alone. Interactions among DMI fungicides and tank-mix partners for CLS severity were observed. In addition to the field trial, single-conidia isolates were collected, representing each treatment before and after fungicide applications. The genetic diversity of isolates was evaluated with 8 fluorescent-tagged microsatellite markers. An estimated 306 original multilocus genotypes (MLGs) were observed among 763 isolates, with 93 MLGs identified before fungicide applications and 250 MLGs identified following exposure to fungicide treatments. Interestingly, only 37 MLGs were observed in both collection periods. The results from this field study will help formulate effective and practical fungicide-based recommendations for managing CLS and understand the impact of fungicides on *C. beticola* genetic diversity.

LONGERSTAY PIERRE*, HEINRICH REINEKE [2], JAN SELS [1], JUAN VEGAS[1], ERIK DE BRUYNE[1], DIRK HYNDRIKX[1], NIELS WYNANT[1] and HENDRIK TSCHOEP[1], [1]SESVanderHave Belgium, Industriepark 15 Soldatenplein Z2, B – 3300 Tienen, [2]SESVanderHave Deutschland GmbH, Am Schloß 3, D – 97084 Würzburg.

Breeding for SBR tolerant varieties at SESVanderHave.

SBR (Syndrôme des Basses Richesses) is a sugar beet disease caused by the proteobacterium *Candidatus Arsenophonus phytopathogenicus* and the phytoplasma *Candidatus Phytoplasma solani*, both transmitted by the planthopper *Pentastiridius leporinus*. The economic impact of the disease on sugar beet can be very significant, as it involves an absolute 5% decrease of the sugar content and a relative loss of 25% in root yield, threatening the cultivation of sugar beet in regions already confronted to many other biotic and abiotic stresses. Since its identification in France in the early nineties, the disease as spread to more than 30,000 ha across Germany and Switzerland and is believed to spread further towards regions neighboring the affected areas. Because of the complex life cycle of the vector insect, agronomic and chemical solutions show a low effectiveness. Hence, resistant varieties are seen as the best mean to ensure a successful cultivation in infested areas. Since years, SESVanderHave breeders have been working on the identification of tolerant genotypes, leading to the release of two varieties in Germany with superior performance in SBR infested conditions, offering for the first time a sustainable solution to farmers affected by the disease. We are now expanding our research efforts to combine our solution against SBR with other desired traits and offer more varieties adapted to the whole sector needs. Moreover, SESVanderHave closely monitors the spread of the disease and is helping farmers to identify the presence of SBR in their fields.

MAJUMDAR, RAJTILAK*[1], PAUL J. GALEWSKI[12], CARL A. STRAUSBAUGH[1] and IMAD EUJAYL[1], USDA-ARS, Northwest Irrigation and Soils Research Laboratory, USDA-ARS, Kimberly, ID, 83341, USA, [2]Plant Germplasm Introduction and Testing Research Unit, USDA-ARS, Pullman, WA, 99164, USA (current).

Beet curly top virus strain specific interactions with sugar beet genotypes reveal potential role of viral small non-coding RNAs (sncRNAs) and small peptides during early infection.

Sugar beet is an economically important crop in the United States (U.S.) and Europe and is highly susceptible to the *Beet curly top virus* (BCTV) which negatively affects sugar beet yield and sugar production. In the western U.S., the predominant strains of BCTV in sugar beet include CA/Logan, Colorado, Severe, and Worland. Severity of disease symptoms depend upon different factors, including virus strain/s, plant genotype, age at infection. The regulatory role of BCTV strain specific small non-coding RNAs (sncRNAs) and their interaction with the host is unclear. Using BCTV susceptible and resistant genotypes, natural infection, and global RNAseq, we demonstrate differential regulation of sugar beet genes by BCTV strain specific sncRNAs. Among detected sncRNAs, sncRNA_36 as an example was common to all four strains and showed moderate negative correlation with *EL10Ac9g22982* (transmembrane protein 53) gene expression in the susceptible line. Whereas *CA/Logan* specific sncRNAs namely sncRNA_4, 20, 21 showed higher negative correlation with *EL10Ac1g01206* (LRR protein), *EL10Ac5g12605* (*7-deoxyloganetic acid glucosyltransferase*), and *EL10Ac6g14074* (*transmembrane emp24*) expression in the susceptible line respectively. Peptidomics analysis identified putative small open reading frame (sORF) derived peptides from BCTV strains. The data presented here suggest that genome divergence among BCTV strains differentially affects the production of sncRNAs and small peptides which could potentially affect pathogenicity and disease symptom development.

REKOSKE, M. M.*[1], O'BOYLE, P.D.[1], LEIN J.C.[2], [1]KWS Seeds, LLC. 1325 Valley View Road, Shakopee, MN 55379 and [2]KWS SAAT AG, Grimsehlstrasse 31, 37555 Einbeck, Germany.

Stacking the deck against Cercospora Leaf Spot.

Cercospora leaf spot (CLS) of sugar beet caused by the fungus *Cercospora beticola* Sacc. is a disease of economic importance across most the North American sugar beet production acreage. The integrated disease management strategy in many markets is the use of fungicides combined with highly genetically tolerant hybrids. While historically there has been an inverse relationship between genetic tolerance to CLS and yield performance in the absence of the disease, new CR+®™ hybrids provide statistically significant advancements (LSD 0.05% as determined either by Official Variety Trial and/or proprietary data). CR+®™ hybrids are composed from traditional multi-genic quantitative resistance sources that exceed CLS approval criteria stacked with newly identified sources of resistance. Breeding gains include significantly greater CLS tolerance based on the KWS 1-9 rating scale, without performance loss in the absence of CLS and provide a substantial performance advantage in terms of sugar per acre under conditions favorable for severe CLS. The benefit of the novel CR+®™ hybrids is due to the balanced composition of different sources of CLS protection combined with traits such as rhizomania, nematode, curly top, root disease and herbicide tolerance that maintain high yield and sugar content under many different environmental situations. The goal of KWS is to further advance sugarbeet as a competitive and sustainable crop and to deliver new innovative solutions for the continued improvement of CLS tolerance in robust and well adapted hybrids.

METZ, NICHOLAS*[1], DORN, KEVIN [1] and KUMAR, AVNEESH [2], [1]USDA-ARS, 1701 Centre Ave, Fort Collins, CO 80526 [2]Colorado State University, 1701 Centre Ave, Fort Collins, CO 80526.

Mapping Rhizoctonia Crown and Root Rot resistance from FC709-3 using new genomic resources.

Rhizoctonia solani is the fungal pathogen that causes Rhizoctonia Crown and Root Rot (RCRR) on sugar beet (*Beta vulgaris*). This devastating disease generates millions of dollars of yield loss throughout most sugar beet growing regions. Tapping decades of RCRR prebreeding lines from the USDA-ARS Fort Collins, CO, we have selected the most resistant accession, FC709-2. This accession was further enhanced for RCRR resistance using three additional cycles of mass selection followed by the selection of a highly resistant individual plant which was used for single seed descent (FC709-3). A single FC709-3 plant was selected for PacBio HiFi and Dovetail Omni-C sequencing to generate a highly contiguous de novo genome assembly. Using this purpose-built chromosome scale genome assembly, we aim to map the genetic underpinnings of RCRR resistance in FC709-3 using RNA-seq, QTL mapping and bulk segregant analysis.

RAMACHANDRAN, VANITHARANI*[1], CHINNARAJA CHINNADURAI[2], NATHAN WYATT[1], ERIC RIVERA SANTIAGO[1], ALYSSA FLOBINUS [2], OLIVER T NEHER[3], JOHN WEILAND[1], MOHAMED KHAN[2] and MELVIN BOLTON[1], [1]United States Department of Agriculture - Agricultural Research Service, 1616 Albrecht Blvd. N, Fargo, ND 58102, [2]Department of Plant Pathology, North Dakota State University, 306 Walster Hall, 1402 Albrecht Blvd, Fargo, ND 58102, [3]Amalgamated Sugar Company, 1951 S. Saturn Way, Boise, Idaho 83709.

High throughput sequencing-aided molecular characterization of resistance-breaking variants of Beet necrotic yellow vein virus in the United States.

Rhizomania, caused by *Beet necrotic yellow vein virus* (BNYVV), is an economically important disease of sugar beet industry globally. The disease is primarily managed by genetic resistance conferred by *Rz1* and *Rz2* genes. However, due to consistent disease pressure, the resistance has been overcome by resistance-breaking (RB) isolates worldwide. To characterize the molecular changes that occurred in the BNYVV genome that confer resistance breaking, we collected rhizomania-infested soil samples from sugarbeet production fields of California, Idaho, Minnesota, and North Dakota that were suspicious for RB ability. To recover the virus from the soil samples, soil-baiting was performed by growing sugar beet varieties possessing no known resistance to rhizomania (susceptible), with *Rz1* alone, and with a combination of *Rz1+Rz2* resistance genes in the test soils. Enzyme-linked immunosorbent assay (ELISA) was used to confirm the presence of BNYVV in the roots of the bait plants from susceptible, *Rz1*, and *Rz1+Rz2* varieties. To gain insight at the nucleotide sequence level, high throughput sequencing (HTS) was applied to the total RNA isolated from the roots of *Rz1+Rz2* bait plants to characterize the molecular changes associated with potential RB-strains. Among the various components of BNYVV genome, detailed analysis of the deduced amino acid sequence of P25 gene, which was previously implicated for RB pathotype, showed mutations at positions 67 and 68 on the hypervariable 'tetrad' amino acids (67-70). In addition, we observed scattered mutations in other area of the P25 gene. This study enhances our current understanding of the molecular changes that are associated with the developing RB population of BNYVV and underscoring the importance of alternative disease management strategies for the future.

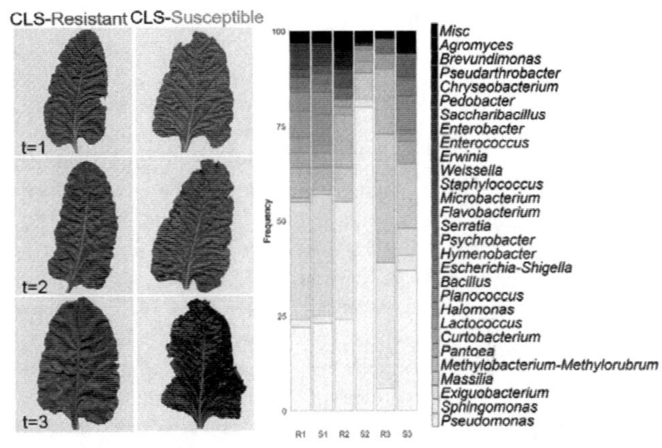

RANGEL, LORENA I.*[1], NATHAN WYATT[1], MARI B. NATWICK[2], PETER C. HAKK[2], MOHAMED F.R. KHAN[2] and MELVIN D. BOLTON[1], [1]USDA-ARS, 1616 Albrecht Blvd. N Fargo, ND 58102-2765, [2]North Dakota State University, Department of Plant Pathology, NDSU Dept 7660 PO Box 6050 Fargo, ND 58108.

Bacterial phyllosphere communities diverge upon invasion of the Cercospora leaf spot (CLS) pathogen in CLS resistant and susceptible sugar beet.

Cercospora leaf spot (CLS), caused by the fungal pathogen *Cercospora beticola*, is the most destructive foliar pathogen of sugar beet (*Beta vulgaris*). Located on the Minnesota and North Dakota border, the Red River Valley (RRV) is the top growing region for sugar beet in the United States. Sugar losses due to CLS in the RRV can be devastating as a result of foliar destruction by this fungus and subsequent biomass loss of the sugar beet root. We sampled leaves of CLS-resistant and -susceptible cultivars grown in a single field in the RRV to monitor changes in the phyllosphere microbiome over the growing season. CLS-resistant and -susceptible sugar beet varieties were allowed to naturally acquire CLS. Leaves from both genotypes were harvested every three weeks beginning at symptom development of the CLS-susceptible variety and ended during harvest. The full 16S rDNA gene was sequenced for each sample to characterize the leaf bacterial microbiome. Bacterial communities were profiled, and composition and predicted functions were compared between sugar beet genotypes. By monitoring communities for taxa associated with the presence or absence of CLS, we can make associations that can inform better management of CLS in an ever-changing environment.

ROSSI, VALENTINA*[1], LOUISE HOLMQUIST[1], ERIK ALEXANDERSSON[2] and LAURA GRENVILLE-BRIGGS[2], [1]DLF Beet Seed, Säbyholmsvägen 24, S-26191 Landskrona, Sweden, [2]Department of Plant Protection Biology, Swedish University of Agricultural Sciences, Box 190, SE-234 22 Lomma, Sweden.

Transcriptome analysis of sugar beet in response to the pathogenic oomycete *Aphanomyces cochlioides.*

Sugar beet (*Beta vulgaris* L.) is one of the main sources of sucrose, providing nearly 30% of sugar production worldwide. The viability of the crop is threatened by the attack of pathogens that cause various diseases, resulting in severe yield losses. The oomycete *Aphanomyces cochlioides* is one of the most problematic pathogens, due to its worldwide distribution and the ability to induce infection at any stage of sugar beet lifecycle, causing seedling damping-off and chronic root rot. During the early stages of sugar beet cultivation the infection can be controlled by the application of chemical fungicides on seeds. However, no major control measures are available for the disease management on mature roots. The genetic basis of resistance is still unclear, therefore the identification of genes associated with sugar beet defense responses is an important step for sustainable disease control. In this study, we performed a transcriptome analysis of partially resistant and susceptible sugar beet breeding lines infected with *A. cochlioides.* Differential expression analysis combined with Gene Ontology (GO) enrichment analysis revealed changes in genes associated to defense mechanisms during the early stages of infection. GO categories associated with hydrogen peroxide (H_2O_2) metabolism, detoxification and cell wall organization were significantly enriched in the differentially expressed gene set from the two partially resistant lines. The findings of this study shed light on the biological processes activated in response to *A. cochlioides* and can be used to assist sugar beet breeding in developing resistant varieties.

RUDE, SAMANTHA*[1], CORY HIRSCH[1] and ASHOK CHANDA[2], [1]Department of Plant Pathology, University of Minnesota, 1991 Upper Buford Cir, 495 Borlaug Hall, St Paul, MN 55108, [2]Dept. of Plant Pathology & Northwest Research and Outreach Center, University of Minnesota, U of M NWROC, 2900 University Ave., Crookston, MN 56716.

Optimizing the screening of sugarbeet lines for resistance to Aphanomyces seedling damping-off.

The soilborne oomycete *Aphanomyces cochlioides* is the causal agent of Aphanomyces seedling damping-off and root rot, important diseases of sugarbeet in Minnesota and North Dakota as well as other sugarbeet growing regions of the world. Various chemical and cultural control measures can reduce disease severity; however, losses due to infection can still be severe under favorable environmental conditions. Adult plant resistance can be effective in limiting losses due to disease, but the genetic mechanism underlying resistance is largely unknown. Screening for resistance is typically done in naturally infested fields or established field disease nurseries. Field-based screening is not effective for phenotyping individual plants and also disease pressure is not uniform across a given field. Information on individual plants will facilitate the identification of regions of the sugarbeet genome associated with resistance to Aphanomyces. Two inoculation types (mycelium and zoospores) were tested for screening sugarbeet lines for resistance to seedling damping-off. Two *A. cochlioides* isolates, both originated from disease nurseries, were used to determine whether or not the optimal inoculation procedure is different depending on the isolate used. For both isolates, inoculation by mycelial suspension resulted in less disease, as measured by root rot index values, compared to inoculation by zoospores. Additionally, disease severity varied between experiment replications to a greater degree in mycelial inoculations than those inoculated by zoospores. In order to identify the ideal inoculum concentration that allowed for discriminating (phenotypic) resistance rating of sugarbeet lines, four zoospore concentrations were compared. Disease severity showed no significant differences based on inoculum concentration, as the majority of plants were dead by 14 days post-inoculation at all concentrations, whereas disease progress, measured by the change in percentage of living plants compared to initial stand, was statistically different between concentrations. The zoospore concentration that led to sufficient symptom development in susceptible plants was found to be at 5,000 zoospores per pot. Taken together, these experiments suggest that zoospore inoculation (particularly at 5,000 per pot) can be used for phenotyping sugarbeet germplasm for resistance to Aphanomyces seedling damping-off.

SECOR, GARY*[1], VIVIANA, RIVERA[1], NATHAN WYATT[2], JUDITH RENGIFO[1] and MELVIN BOLTON[2], [1]North Dakota State University, Plant Pathology, Fargo, ND 58102, USDA-ARS Schafer Agricultural Research Center, Fargo, ND 58102.

Early *Cercospora beticola* detection and germination in commercial sugarbeet fields.

Cercospora leaf spot (CLS) cause by *Cercospora beticola* (Cb) continues to be an important sugarbeet disease in all production areas in the world. Most CLS forecasting models predict conditions favorable for disease spread and application of fungicides after disease is present, but do not include conditions influencing spore production and germination that may influence early fungicide application before infection by Cb. Because we observed spore germination in Cb cultures stored at 10C, we conducted laboratory experiments to determine how temperature and moisture affect spore germination. Across all treatments, spore germination is higher in free water (95%) compared to 100% RH (30%). Germination begins in two hours at 10°C and increases with time and temperature. Across all treatments, spores from fungicide resistant isolates have a lower percent germination compared to fungicide sensitive isolates, and significantly lower percent spore germination at temperatures <14°C but not at >18°C. This implies there may be fitness penalty for spore germination of resistant isolates at colder temperatures that disappears at warmer temperatures. Preliminary studies in sugar beet fields in ND and MN show that earlier fungicide applications improves CLS control. Field experiments in 2021 (a dry spring) and 2022 (a wet spring) to were conducted to study timing of Cb spore detection and infection using spore traps (Spornado) and weather stations with leaf wetness sensors in six commercial sugarbeet fields. Filter cartridges collected three times per week from early May to early August were tested for the presence of Cb DNA by PCR. From early June to late July of 2021, asymptomatic leaves from 57 commercial fields were tested for the presence of Cb DNA by PCR. Some sites were monitored for appearance of CLS spots. In 2021, first Cb spores detection in some field traps was May 3, Cb DNA first detected in asymptomatic leaves June 14 and 17, and CLS spots first observed June 29. In 2022, a wet time, Cb first spore detection was May 23 in all spore traps and continued in all locations until completion of trapping on July 8. Spore DNA was tested for sensitivity/resistance to QoI and DMI fungicides. Based on our studies, we conclude that forecasting models for CLS should include spore detection and early weather conditions, and adjusted to recommend fungicide applications earlier in the growing season before infection by Cb. This study also confirms the utility of Cb spore trapping and PCR detection of infection by Cb before CLS is present.

SNELL, PER*, DLF Beet Seed, Säbyholmsvägen 24, 26194 Landskrona.

Novel genetic sources paves the way for sustainable Cercospora resistance in sugar beet.

Cercospora leaf spot disease (CLS), caused by the hemibiotrophic fungus *Cercospora beticola,* is one of the most severe sugar beet diseases causing a significant yearly loss in total sugar yield in the US. Identifying genetic resistance has long been an important goal for sugar beet breeding companies and has resulted in two types of genetic resistance; monogenic and multigenic. Experience from other monogenic resistance traits shows that monogenic resistance is easily overcome by the pathogen by drastically changing the selection pressure. In order to achieve sustainable CLS resistance for the US market multigenic is required. In this work, we present the genetic background of several novel multigenic resistance sources and a strategy for introducing a genetic defence-in-depth approach for long term resistance against CLS in the US sugar beet market.

TODD, OLIVIA*, DORN, KEVIN, and HANSON, LINDA. USDA-ARS, Crops Research Laboratory, 1701 Centre Ave, Fort Collins, CO 80526.

Standard area diagram development for *Fusarium oxysporum* symptom and disease progression rating in *Beta vulgaris.*

Standard area diagrams (SADs) are disease progression diagrams that have both a written and visual component designed to help a rater estimate plant disease severity in a uniformly repeatable way. The goal of our study was to record the disease ratings given by volunteers with varying experience rating *fusarium oxysporum* infected sugar beets. Participants rated the same set of images twice, first using a previously published 1-5 scale and again using the standard area diagram. The new *fusarium oxysporum* sugar beet disease progression SAD uses a percentage scale that accounts for leaf wilt, yellowing and necrosis. Use of a SAD was proposed to improve the accuracy of raters compared to a previously published scale. The efficiency of sugar beet breeding programs depends on accuracy of the screening methods. Thus, an improved and replicable rating scale based on a standard area diagram will benefit both public and private breeding efforts and ultimately sugar beet growers. This updated tool will ensure that the USDA-ARS Fort Collins pre-breeding program is properly and thoroughly screening accessions to get the strongest and most stable *Fusarium oxysporum* resistance for sugar beet production.

WYATT, NATHAN A.*[1], VIVIANA RIVERA[2], GARY SECOR[2] and MELVIN D. BOLTON[1], [1]Sugarbeet and Potato Research Unit, Edward T. Schafer Agricultural Research Center, 1616 Albrecht Blvd, NCSL SBPR, Fargo, ND 58102, [2]Plant Pathology Department, North Dakota State University, NDSU Dept 7660, PO Box 6050 Fargo, ND 58108-6050.

Detection of latent Cercospora leaf spot infection in commercial sugarbeet fields.

Cercospora beticola, causal agent of Cercospora leaf spot (CLS), is the most economically damaging pathogen of sugar beet (*Beta vulgaris* subsp. *vulgaris)*. Management of CLS is reliant on timely fungicide applications and CLS tolerant sugar beet varieties. In response to fungicide applications, *C. beticola* populations evolved resistance, with documented resistant isolates identified for all major classes of fungicides except for the ethylene bisdithiocarbamate class (EBDC) of fungicides. The economic importance of fungicide resistance cannot be overstated, with CLS epidemics occurring in the upper mid-west in three of four years from 2016 to 2020. To discourage fungicide resistance from developing, disease forecasting models are used to inform growers of the best time to apply fungicides. These models indicate that fungicides should be applied when symptoms of CLS first appear or when environmental conditions are favorable for CLS symptom development. However, the lifecycle of *C. beticola* includes a long latency period of infection where little to no symptoms of infection are observed. To better understand the temporal infection dynamics of CLS during this important early infection stage, we undertook a large cooperative project spanning the full RRV growing region with the goal of identifying latent infection in commercially grown fields. For five weeks in June and July, 280 field samples were collected each week. Each sample was screened for the presence of *C. beticola* using a qPCR assay. Additional, qPCR assays for fungicide resistance to systemic fungicides were simultaneously done to detect resistance to benzimidazoles, triazoles (DMIs), and quinone outside inhibitors (QoIs). Results from this study provide the first step in developing more precise timelines of CLS disease progression and information that should be incorporated into CLS disease forecasting models.

WYNANT NIELS*[1], JAN SELS[1], JUAN VEGAS[1], ERIK DE BRUYNE[1], MARK ANFINRUD[1] and HENDRIK TSCHOEP[1], SESVanderHave, Industriepark 15, 3300 Tienen (BELGIUM)

Breeding of insect tolerant varieties at SESVanderHave.

Various insect tolerances are becoming increasingly important for trait stacking in modern sugar beet varieties, as more and more chemical solutions are being banned, either as field or seed treatments. SESVanderHave has dedicated breeding efforts on insect tolerances and on the diseases vectored by insects such as SBR, aphid resistances, virus yellows, beet moth, curly top, and root maggots, amongst others.This presentation will focus on breeding efforts done for the virus yellow complex, as these viruses are spread by aphids. It is a complex of viruses including the Beet Yellows Virus (BYV), the beet mild yellowing virus (BMYV) and the beet Chlorosis Virus (BChV). In the 20th century these viruses were considered as one of the worst threats to sugar beet cultivation in north-west Europe. After the restrictions on neonicotinoid-based insecticide and milder winters and springs because of climate change, these viruses made a dramatic come-back in 2020, resulting in yield losses of between 0-40% across Western Europe. SESVanderHave identified "Virus Yellows" as a top priority in its research and development program as early as 2010. Since then, we have increased our investment in disease phenotyping platforms, scientific expertise, molecular diagnostics tools and used our diverse germplasm collection to enable large-scale testing for a genetic solution to BYV, BMYV and BChV. These investments in resources and time have enabled us to exploit the genetic diversity to identify VY-tolerant germplasm to create the first generation of VY-tolerant varieties. In addition, relations to other insect tolerance breeding programs, such as root maggot tolerance, will be elaborated on. ☐

POSTER PRESENTATIONS

BEYER, WERNER*[1], JENNIFER PETERSON[2], HARALD KEUNECKE[1] and MARGARET REKOSKE[2], [1]KWS SAAT SE & Co. KGaA, Grimsehlstrasse 31, 37574 Einbeck, Germany, [2]KWS SEEDS, LLC, 1325 Valley View Rd, Shakopee, MN 55379.

Early Aphanomyces resistance as key element for future robust sugarbeet varieties.

Aphanomyces cochlioides Drechs. is a soil-borne oomycete fungus causing both an acute seedling disease (damping-off) as well as a chronic rot of the mature root (black root) in sugarbeet. The pathogen can lead to serious damage in most sugar beet growing areas, including North America, Europe and Asia. Chemicals like Hymexazole (Tachigaren®) have been instrumental in protecting the early developmental stages against seedling death and protect the emerging plantlets from early infection, which contributes to heavy late root rot. While the late phase of Aphanomyces root rot has been addressed by breeding companies for a long time, genotypic variation for early Aphanomyces infection was of lower priority in part due to highly effective chemical protectants. Various sources of genotypic variability for early Aphanomyces have been reported, i.e., from wild beet resources or USDA germplasm. Within KWS, we have been working on early Aphanomyces resistance over the past years and are establishing routines for larger germplasm screenings. Current genetic variation is probably not strong enough to replace existing chemical protection but improvements would provide complementary safeguarding. The goal of our breeding approach is to integrate early Aphanomyces resistance into variety development to further strengthen the defense line against soil-borne diseases. Along with an overall improved root and leaf healthiness and disease resistance packages this shall serve as a key element for future robust sugar beet varieties.

BUBLITZ, DANIEL* and THOMAS WENZEL, Sugarbeet Advancement, Michigan State University, 3775 S Reese Rd, Frankenmuth, MI 48734.

Re-evaluating the management of *Rhizoctonia solani* on sugarbeet in Michigan through the use of variety tolerance and Quadris applications.

Rhizoctonia root and crown rot, caused by the fungal pathogen *Rhizoctonia solani*, is a major soilborne disease of sugarbeet in Michigan. This disease has caused substantial losses to growers, both in terms of total root yield and sugar concentration. In Michigan, severe infections have been observed to cause up to a 15-ton reduction in yield and a 1 percentage point decrease in sugar content. Throughout the region, this disease is primarily managed by planting tolerant sugarbeet varieties and the use of fungicide applications. The most common fungicide used is Quadris (Azoxystrobin), which many growers apply twice during the season, with the first application T-band in-furrow at planting and the second in a 7-inch band at the 6-10 leaf stage. The combination of these management tools has been effective in managing Rhizoctonia in Michigan. As sugarbeet varieties continue to improve, it has been observed that variety tolerance to Rhizoctonia is stronger in many contemporary varieties than in many varieties available when these recommendations were first made. With this increased variety tolerance, it is hypothesized that two applications of Quadris may no longer be necessary in the majority of the Michigan beet production region. Therefore, the current study compared a Rhizoctonia susceptible variety and a tolerant variety, each with four combinations of Quadris applications, including no Quadris applied, one application in-furrow, one application at the 6 to 10 leaf stage, and two applications. This study was conducted over the course of three years, being done in a different non-inoculated grower field each year. In 2020 and 2021, the fields had high levels of Rhizoctonia, while the field in 2022 had a moderate level of disease. To evaluate the impact of the different treatments, the number of dead or dying beets were counted in 1200 feet of row, and yield and sugar data were collected. Preliminary results from the first two years have shown that for a tolerant variety in a field with high levels of disease, there was no difference between a single application or two applications of Quadris. For the susceptible variety in this environment, two applications of Quadris provided greater control than either single application, all three of which provided greater control than no applications of Quadris. If the number of Quadris applications can be reduced in certain situations, this would provide cost savings to growers and reduce the risk of *Rhizoctonia solani* developing resistance to Quadris.

COURNEYA, ISAAC*[1], OLIVIA HAMILTON[1], LORENA RANGEL[2] and MELVIN D. BOLTON[2], [1]North Dakota State University, NDSU Dept 7660, PO Box 6050 Fargo, ND 58108, [2]USDA - ARS, 1616 Albrecht Blvd N, Fargo, ND 58102.

Characterization of a novel effector protein in *Cercospora beticola*.

Cercospora leaf spot (CLS), caused by the fungal pathogen *Cercospora beticola*, is the most economically important disease of sugar beets worldwide. Leaf spots formed by the pathogen coalesce to form large necrotic regions which can kill entire leaves. The regrowing of leaves necessary to maintain photosynthesis comes at the expense of sugar stored in the root, which reduces yield. In order to combat this disease, it is important to understand how infection and necrosis are perpetrated at a molecular level. *C. beticola* uses an array of effector molecules, including proteins and secondary metabolites, to cause disease. While a few *C. beticola* effectors have been characterized, there are many more effectors yet to be identified. Recent culture filtrate analysis has suggested the existence of multiple unidentified potential protein effectors. Among these candidate effector proteins is gene 05663, which has been predicted to be an apoplastic effector due to its small size and secretion signal. Using PEG-mediated transformation, we have successfully created five Δ05663 mutants. By comparison to the wild-type strain and three ectopic strains, Δ05663 mutants present delayed symptoms in inoculated sugar beets, suggesting that gene 05663 plays an integral role in the disease-causing process. Inoculation results, protein characterization data, and the potential role this gene plays in virulence will be presented.

CZARNECKI, OLAF*, WERNER BEYER and JENS LEIN, KWS SAAT SE & Co. KGaA, Grimsehlstrasse 31, 37574 Einbeck, Germany.

Syndrome des Basses Richesses - A novel insect transmitted disease in sugar beet.

Climate change has hit sugar beet farmers in certain regions of Southern and Eastern Germany and Switzerland with full power. Surprisingly, not only drought related abiotic stress has become a major challenge in these regions but also occurrence of a plant hopper (*Pentastiridius leporinus*) transmitting a bacterial disease to sugar beet. Plants infected with bacterial species *Candidatus Arsenophonus phytopathogenicus*, develop a disease called "Syndrome des Basses Richesses" (SBR). SBR infected beet show serious symptoms including leaf yellowing and necrosis, growing of asymmetric, lancet-shaped leaves, browning of tap root tissue but most importantly loss of sugar content up to 5% (abs.) and yield losses up to 25%. With declining sugar prices in SBR affected regions farmers almost fail to keep sugar beet as valuable and profitable part of their crop rotation threatening existence of regional sugar factories. Until now, the plant hopper development cannot be controlled by means of chemical or agronomical plant protection laying a focus on genetic solutions. For SBR resistance breeding KWS has setup various field test locations in Southern and Eastern Germany and a greenhouse test system to examine genotypic variation for effects of SBR infection on sugar yield. A broad range of experimental hybrids is continuously tested for their performance in infected growing areas which will provide KWS varieties with stable and high yields upon SBR infestation in short-term. For long-term solutions and to ensure future breeding gain, we have established a new breeding program to screen for and make use of novel genetic resistance sources against SBR. The program employs most recent means of modern plant breeding including fast breeding cycles, advanced phenotyping, genomic data analysis and holistic variety development. KWS is working with highest priority on providing SBR tolerant sugar beet varieties in the shortest possible time to preserve sugar beet in SBR infested regions.

DERVARIC CHRISTINE*[1], LINDA HANSON[2], MARY RUTH MCDONALD[3] and CHERYL TRUEMAN[1], [1]University of Guelph, Ridgetown Campus, 120 Main Street East, Ridgetown, ON N0P 2C0, Canada, [2]USDA-ARS, 612 Wilson Road, East Lansing, MI 48824, [3]University of Guelph, 50 Stone Road East, Guelph, ON N1G 2W1, Canada.

Do fungicides containing phosphites have a direct effect on *Cercospora beticola* Sacc., the causal agent of Cercospora leaf spot of sugarbeet?

Cercospora beticola Sacc., causal agent of Cercospora leaf spot (CLS), is one of the most serious foliar pathogens of sugarbeets (*Beta vulgaris* spp. *vulgaris*). Resistance to fungicide resistance action committee (FRAC) Group 1 and 11 fungicides and increasing insensitivity to Group 3 fungicides provide increasing challenges for CLS management in Ontario, Canada. The active ingredient mancozeb, has been an important fungicide in standard spray programs, but has been re-evaluated with a reduction in maximum applications from eight to five permitted. Evaluation of alternative fungicides in 2019, 2020, and 2021 field trials demonstrated that fungicides containing phosphites are a possible alternative to mancozeb, providing a significant reduction in disease through-out the season in all three field trials. To use these products effectively, it is beneficial to know if they directly suppress the pathogen or have an indirect effect such as increasing host resistance. Five concentrations of a commercial formulation of phosphites (Phostrol) (0.01, 0.1, 1, 10, and 100ul/ml agar) were tested for effect on mycelial growth of eight single spore isolates of *Cercospora beticola*. Single spore isolates were collected from sugarbeet leaves with CLS symptoms from the outer guard rows of research field trials in 2020, and 2021 in Ridgetown, Ontario, Canada. Phosphites were incorporated into the half V8 agar media, a mycelial plug placed in the center and the diameter of cultures was measured after five days. No concentration of phosphites decreased mycelial growth in comparison to the non-treated control (p-value less than 0.05). The 100ul phosphites per ml treatment (highest concentration tested, mean 18.4 mm) was statistically equivalent to that of the non-fungicide control (mean 17.6 mm) when analyzed using SAS 9.4 PROC GLIMMIX. This supports the hypothesis that phosphites stimulate host resistance to the pathogen and does not directly suppress the pathogen. Concurrent research on fungicide programs integrating phosphites is expected to provide growers with another option to combat CLS.

DORN, KEVIN*[1], NIELSON, AMY[1], FENWICK, ANN[1], KUMAR, AVNEESH[2] and MCGRATH, MITCH[1], [1]USDA-ARS, Crops Research Laboratory, 1701 Centre Ave, Fort Collins CO, 80526, [2]Colorado State University, Crops Research Laboratory, 1701 Centre Ave, Fort Collins CO, 80526.

15 trillion base pairs of beet DNA sequencing enables a genomic retrospective of 8 decades of breeding progress.

The post-genomics era of sugar beet trait discovery and breeding is upon us, with the deployment of multiple high quality reference genome assemblies for sugar beet, other beet crop types, and crop wild relatives. Reference genomes like EL10 and RefBeet provide a scaffold on which to build a genomic understanding of the history of sugar beet domestication and improvement. In this study, we present population level whole genome resequencing of a broad panel of 196 beet germplasm spanning an 86 year history of public improvement efforts. The germplasm panel represents early public sugar beet cultivars, USDA-ARS pre-breeding lines, as well as accessions from table beet, fodder beet, and chard. DNA samples of 25 individual plants per accession were pooled together and sequenced to a target of 80 fold genomic coverage (average 75.9 gigabases per sample, minimum = 49.3 gigabases, maximum = 157 gigabases). Utilizing the new, highly contiguous and well-annotated reference genome EL10.2.2, various clustering statistics based on allele frequency datasets from the mapped pools help characterize the genomic history according to crop type and breeding location. Using sliding windows across the genome, diversity statistics such as Fst and Pi identify regions of the genome that differentiate crop types, disease resistance breeding targets, and breeding locations. Collectively, these genome resequencing resources represent a component of a tractable rare allele discovery pipeline for genes of interest. We envision that the integration of these resequencing datasets into BeetBase will provide the beet genetics community a new tool for mining germplasm for useful genetic diversity.

EIDE, JOHN D.*[1], ABBAS M. LAFTA[2], KAREN K. FUGATE[1], and MOHAMED F. R. KHAN[2], [1]USDA-ARS, ETSARC, 1616 Albrecht Blvd. N., Fargo, ND 58102, [2]North Dakota State University and University of Minnesota, NDSU Dept. 7660, PO Box 6050, Fargo, ND 58108.

Drought stress prior to harvest affects postharvest respiration rate and susceptibility to storage rots.

Sugarbeets in North Dakota, Minnesota and Michigan are largely grown without supplemental irrigation. Roots harvested from these production regions, therefore, are inevitably drought stressed when natural rainfall is insufficient. Roots from drought stressed plants are currently incorporated into storage piles despite a paucity of information regarding the impact of water stress on root storability. Research, therefore, was carried out to assist storage pile managers in formulating best practices for the postharvest storage of drought-stressed roots by determining the effects of preharvest drought stress on postharvest respiration rate, sucrose loss, invert sugar accumulation and susceptibility to storage rots. Drought stressed roots were obtained by discontinuing watering of greenhouse plants for 7, 14 or 21 days prior to harvest. These watering regimes yielded roots with mild, moderate and severe water stress as determined by their effects on plant photosynthesis and root water content. Storage respiration rate was significantly elevated by severe water stress, with the magnitude of drought-induced elevations in respiration rate increasing as storage duration increased. Moderate and severe water stress additionally increased susceptibility to rot caused by *Botrytis cinerea* and *Penicillium* spp., but had no significant effect on root sucrose or invert sugar concentrations. Overall, these results suggest that sugarbeet root storability is minimally affected by mild drought stress but is progressively eroded as both drought stress severity and storage duration increase.

FENWICK, ANN*[1], AMY NIELSON[1], NICK METZ[1], AVNEESH KUMAR[2], OLIVIA TODD[1] and KEVIN DORN[1], [1]USDA-ARS SMSBRU, 1701 Centre Ave, Fort Collins, CO 80526, [2]Department of Agricultural Biology, CSU, Fort Collins, CO 80521.

Development of a reliable greenhouse screen for Rhizoctonia Crown and Root Rot to identify resistant germplasm and develop a non-destructive rating method.

For over 30 years, the USDA-ARS Crop Germplasm Committee has organized the screening of Beta germplasm for resistance to multiple critical diseases. The CGC effort screens between 30-50 Plant Introductions (PI's) from the USDA National Plant Germplasm System each year for multiple diseases. Fort Collins has consistently conducted the Rhizoctonia Crown and Root Rot (RCRR) resistance screening for over 50 years. The evaluated germplasm results are summarized in the Germplasm Resource Information Network (GRIN) and reported in Plant Disease Management Reports (PDMRs). The Disease Index (DI) reported for each evaluated accession represents a mean of the replicated plot-level DI ratings, which consist of approximately 50 – 75 plants each per accession. In this study, we report a reliable greenhouse (GH) RCRR screening protocol that effectively triples our annual screening capacity. The GH protocol also allows for a non-destructive rating procedure to collect individual DNA and allow resistant plants to survive both the screening process and vernalization. This allows the plants to generate seed for further plant breeding experiments that are very difficult to accomplish with field evaluations alone. This GH protocol has been optimized with sugar beet and now also applied to a core collection of 64 *Beta vulgaris ssp. maritima* (*Bvm*) accessions. The *Bvm* RCRR screen consisted of 1330 plants (1 plant/6" pot) in one large GH bay with 3 large benches. Twenty plants for each of the 63 PI's, (1 PI did not have sufficient germination) and 70 check plants (35 resistant & 35 susceptible) were planted on the same day. After germination, and thinning, all plants of all PI's and checks were randomly positioned in 133 blocks of 10 plants across a large GH bay for the remainder of the experiment. The six *Bvm* PIs with the lowest mean DI and most harvestable individuals (scoring 0-3) are from 4 countries: Denmark (2 PIs), France (2 PI's), Italy and Germany. PI 540677 and PI 540674 both from Denmark, and 1 from France (PI 504276) have 5 plants surviving from each population (n=20). These resistant plants have also survived induction. The top 6 accessions have a mean DI GH score of 20 plants that range from 4.4 to 5.6. The same 6 accessions have previously been evaluated in field trials and have DI scores recorded in GRIN ranging from 3.0 to 6.1. This suggests the GH screen is sufficiently uniform and of an appropriate disease intensity to separate resistant plants from susceptible plants. The GH screen shows uniform pressure over a diverse set of *Bvm* PIs. The reduced GH variability compared to field evaluations that vary year to year, is preferable to making accurate selections in the greenhouse. This new protocol for screening both sugar beet and crops wild relatives provide an additional tool to expedite RCRR resistance trait discovery via germplasm screening, genetic mapping, and marker assisted selection.

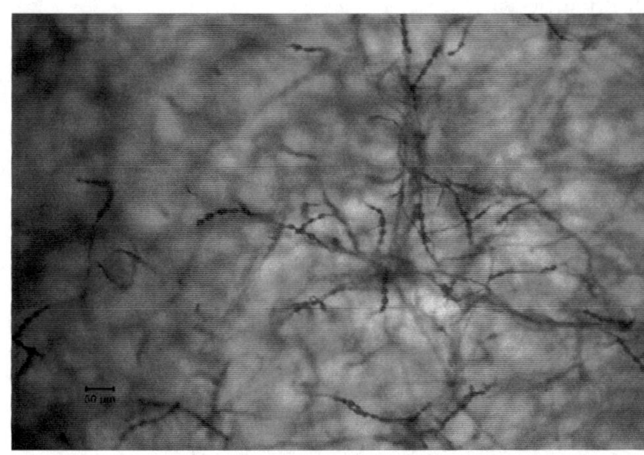

HANSON, LINDA E.*[1], MALINI A. JAYAWARDANA[2] AND DOUGLAS H. MINIER[2], [1]USDA-ARS, 612 Wilson Road, 37 Plant Biology Lab, East Lansing, MI 48824, [2]Michigan State University, 612 Wilson Road, 37 Plant Biology Lab, East Lansing, MI 48824

Diversity of *Alternaria alternata*, causal agent of Alternaria leaf spot of sugar beet.

Alternaria leaf spot of sugar beet has long been a minor issue in the United States. Incidence of this disease has increased in Michigan, with up to 30% of the leaf spot damage in the 2016 through 2019 growing seasons attributed to Alternaria leaf spot. As part of the investigation of factors contributing to this increased disease prevalence, isolates were collected in the field and compared to isolates collected before 2012. Isolates from sugar beet also were compared to strains collected on rotation crops, such as potato and dry bean, and other plants with Alternaria symptoms in the region. All *Alternaria* isolates were morphologically similar to *Alternaria tenuissima*, producing small conidia in largely unbranched chains. However, as recently reported, isolates could not be genetically separated from *A. alternata* (small conidia in branched chains) using a multi-gene sequencing approach. In a three gene phylogeny, isolates formed at least three clades within *A. alternata* and did not cluster with isolates previously suggested as subspecies *tenuissima*. These results agree with previous reports that morphological differences do not reliably relate to genetic variability within *A. alternata*. In addition, isolates recovered prior to 2012 were found in the same genetic groups as those from 2016-2019. Thus, showing no evidence of a substantial change in the fungal population associated with the increase in disease incidence in the region. Isolates originally collected from sugar beet, dry bean, potato, and blueberry all caused lesions on sugar beet leaves with no significant differences in disease severity related to original host. The isolates from sugar beet also grouped with isolates from dry bean, potato, blueberry, and previously published strains from apple and pear, indicating there is no relationship between host and pathogen genotype. Our results support the potential for *A. alternata* from a diverse range of crops to infect sugar beet.

HENDERSHOT, CARLY*[1], CHRIS BLOOMINGDALE[1], HOLLY CORDER[1], TOM GOODWILL[2], SARAH RUTH[1], RANDY BEAUDRY[1], LINDA E. HANSON[2] and JAIME F. WILLBUR[1], [1]Michigan State University, Plant Biology Laboratory, 612 Wilson Rd. Room 35, East Lansing, MI 48824, [2]USDA-ARS, Plant Biology Laboratory, 612 Wilson Rd. Room 37, East Lansing, MI 48824

Evaluation of Cercospora leaf spot on impacts on post-harvest rot susceptibility and storage rot effects on sugar beet respiration.

In Michigan, sugar beets (*Beta vulgaris*) are stored for up to 200 days post-harvest. During storage, sugar content is reduced due to rot and regular energy use from respiration. Cercospora leaf spot (CLS) is hypothesized to be a predisposing factor for storage losses, specifically for increased storage rot. However, previous studies have presented conflicting evidence of this interaction and the effects of CLS remain unclear for specific storage pathogens. To investigate the impact of CLS on fungal storage rot, post-harvest symptom development was evaluated in beets with relatively high or low in-season CLS severity. At three timepoints during storage, roots of each CLS level were inoculated with *Fusarium graminearum, Botrytis cinerea, Penicillium vulpinum,* or *Geotrichum candidum* and symptoms assessed after 7 days. There were no significant differences between storage rot susceptibility to any of the tested pathogens in beets, regardless of CLS level, at any timepoint in any year ($P > 0.05$). The effect of CLS and storage rot inoculation on root respiration rate was also evaluated in respirometry chambers in 2021. Preliminary data shows a significant increase in respiration rate in beets inoculated with pathogens compared to control ($P < 0.05$), as well as variation among respiration rates between the storage pathogens. Studies are ongoing and will increase understanding of factors contributing to potential storage losses. Such knowledge could allow targeted disease management for high impact disease and lead to improved sugar retention and related profit for sugar producers.

HERNANDEZ, ALEXANDRA*[1], DANIEL BUBLITZ[2], TOM WENZEL[2], SARAH RUTH[1], CHRIS BLOOMINGDALE[1], DAVID METTLER[3], MARK BLOOMQUIST[3], LINDA HANSON[1] and JAIME WILLBUR[1], [1]Michigan State University, 612 Wilson Rd, East Lansing MI 48824, [2]Michigan State University Extension and Michigan Sugarbeet Advancement, 3775 S. Reese Road, Frankenmuth, MI 48734, [3]Southern Minnesota Beet Sugar Cooperative, 83550 Co Rd 21, Renville, MN 56284.

An in-field heat treatment to reduce *Cercospora beticola* survival in plant residue and improve Cercospora leaf spot management in sugarbeet.

Sugarbeets account for over 50% of U.S. sugar production. Cercospora leaf spot (CLS), primarily caused by the fungal pathogen *Cercospora beticola*, is a major foliar disease of sugarbeet. Since this pathogen survives in infected leaf tissue, this study evaluated management strategies to reduce inoculum overwintering and survival. Treatments included moldboard plowing immediately post-harvest (6-in. depth), heat treatment with a propane-fueled burner at 1 mph immediately pre-harvest, and application of a desiccant (saflufenacil) 7 days pre-harvest. After treatment, leaf samples were evaluated at 0-, 45-, 90-, and 135-days post-harvest to determine *C. beticola* viability. The following season, inoculum pressure was measured by taking disease ratings on a susceptible beet variety planted into the same plots, and by counting lesions on highly susceptible sentinel beets placed into the field at weekly intervals. The heat treatment significantly reduced lesion sporulation (2019-20 and 2020-21 trial; $P < 0.0001$, 2021-22 trial; $P < 0.05$) and *C. beticola* isolation (2019-20 trial; $P < 0.05$) in at-harvest samples. Reduced numbers of CLS lesions were observed on weekly sentinel beets placed in heat-treated plots from May 26-June 2 ($P < 0.05$) and June 2-9 ($P < 0.01$) in 2019, as well as June 15-22 ($P < 0.01$) in 2020. The heat treatment also reduced the area under the disease progress curve for CLS assessed the season after treatments were applied (2019 and 2020; $P < 0.05$). In 2019 and 2021, a propane-fueled heat treatment of leaf residue in the spring resulted in similar reductions in CLS severity in Minnesota field studies. Overall, heat treatment of fresh or overwintered leaf tissue could be used to aid in CLS management.

HERNANDEZ, ALEXANDRA*[1], SARAH RUTH[1], MIO SATOH-CRUZ[1], DANIEL BUBLITZ[1], LINDA E. HANSON[2] and JAIME F. WILLBUR[3], [1]Michigan State University, 612 Wilson Rd, East Lansing, MI, 48824, [2]USDA-ARS, 612 Wilson Rd, East Lansing, MI, 48824, [3] Michigan State University, 612 Wilson Rd, East Lansing, MI, 48824.

Screening of fungicide resistance in *Cercospora beticola* populations in Michigan using PCR-RFLP and spiral plate gradient methods.

There are multiple fungicide groups that are commonly used and registered for Cercospora leaf spot management in sugar beet including methyl benzimidazole carbamates (MBC or benzimidazole, FRAC group 1), quinone outside inhibitors (QoI or strobilurin, FRAC group 11), demethylation inhibitors (DMI or triazole, FRAC group 3), and multi-site contact activity (FRAC group M03) fungicide classes. In Michigan, reduced sensitivity to QoI, MBC, and DMI fungicides has been detected and extensively monitored in *C. beticola* populations in Michigan. In 2021 and 2022, testing was conducted using polymerase chain reaction restriction fragment length polymorphism (PCR-RFLP) assays to detect point mutations in the *C. beticola* genome associated with fungicide resistance and compared to the effective concentrations to inhibit 50% of mycelial growth (EC_{50}) determined in spiral plate gradient assays. In 2021, 29 field locations were sampled across nine counties in east-central Michigan (n = 418 lesions tested); in 2022, 16 field locations were sampled with submissions continuing at present (n = 223 lesions tested). Resistances were determined by detection of the point mutations present in i) the fungal mitochondrial cytochrome b gene (G143A) for QoI fungicides, ii) in the beta-tubulin gene (E198A) for MBC fungicides, and iii) in the C-14 alpha-demethylase gene (Glu169 (GAA to GAG)) for DMI fungicides. In 2021, high frequencies of QoI resistance were observed across all locations and dates; of the 386 isolates tested, 385 were found to contain the G143A mutation. By the final sampling date of the season, mutations for resistance to QoIs and MBCs were present in 100% of the fields tested, while the mutation conferring high levels of resistance to DMIs was found in 83% of fields. In 2021, *in vitro* screening of corresponding *C. beticola* isolate sensitivities was conducted for eight fungicide active ingredients including pyraclostrobin (PYR), thiophanate-methyl (TPM), difenoconazole (DIF), fenbuconazole (FEN), mefentrifluconazole (MEF), prothioconazole (PRO), tetraconazole (TET), and triphenyltin hydroxide (TPH) (n = 134 isolates). Based on resulting EC_{50} values, the majority of isolates (≥90%) were sensitive to DIF, FEN, MEF, and TPH at concentrations of ≤10 µl ml^{-1}. Elevated frequencies of isolates with EC_{50} values >10 µl ml-1 were identified for PYR (43%), TPM (62%), PRO (69%), and TET (40%). Current year testing is in progress, however, correlations between spiral plate and PCR-RFLP methods will be evaluated for each active ingredient to optimize future screening efforts.

LENNEFORS, BRITT-LOUISE*, LOUISE HOLMQUIST, PER SNELL and JOAKIM HERRSTRÖM, DLF Beet Seed AB, Säbyholmsvägen 24, Landskrona, 261 91, Sweden.

Sugar Beet hybrids with resistance/tolerance to Virus Yellows entering the European market.

After the ban of the neonicotinoides within the EU in 2018, the ability to control aphids in sugar beet production has drastically deceased. This includes the peach potato aphid (*Myzus persicae*), the main vector of viruses causing Virus Yellows. In Europe, Virus Yellows is caused by one or more of beet yellows virus (BYV), beet mild yellowing virus (BMYV) and beet chlorosis virus (BChV). In several European countries Virus Yellows has caused significant yield losses with reports of about 70% yield loss in some growing areas.

At DLF Beet Seed, research activity towards the different viruses as well as identifying and collecting naturally occurring resistance against the disease goes back to the 1960's. Hybrids based on resistance and tolerance to more than one virus have been developed and are on the way to the European market under the name VY-tech® varieties.

LENNEFORS, BRITT-LOUISE*[1], BENGT JEPPSSON[1], ADAM HANSEN[2], TYSON WATERS[2] and TYLER RING[2], [1]DLF Beet Seed AB, Säbyholmsvägen 24, Landskrona, 261 91, Sweden, [2]DLF Beet Seed LLC, 1020 Sugar Mill Road, Longmont, CO 80501.

Screening root maggot lines/hybrids for genetic tolerance.

Sugarbeet Root Maggot, *Tetanops myopaeformis* is causing increased production challenges for several areas of the Sugarbeet growing regions. Northeastern North Dakota, Northwestern Minnesota, Wyoming/Colorado, and Idaho are a few areas where Sugarbeet Root Maggots (SBRM) are present. Root Maggots reduce plant stands, plant vigor and can cause a significant reduction of grower yield and profit. Control of SBRM is becoming more challenging as key insecticides like Chlorpyrifos have been banned. This has caused growers to be become more aggressive and timelier with applications of alternative insecticides that are currently registered for use. In addition to these common production practices, growers are also reliant on Sugarbeet insecticide seed treatments to help with control. DLF Beet Seed recognizes the importance and continued need for SBRM tolerant hybrids and is actively placing emphasis on our SBRM breeding program to address the growing concern of Root Maggots throughout the US. A combination of line development and line evaluation has taken place in the Red River Valley and is showing positive results for improved SBRM tolerance in DLF Beet Seed varieties. We plan to continue these research efforts to bring our growers the best hybrids possible for years to come.

LIEN, AUSTIN* and ASHOK CHANDA, Dept. of Plant Pathology & Northwest Research and Outreach Center, University of Minnesota, U of M NWROC, 2900 University Avenue, Crookston, MN 56716.

Use of infested sorghum for *Cercospora beticola* inoculation.

Cercospora leaf spot (CLS), caused by *Cercospora beticola* Sacc., is responsible for significant economic loss in sugarbeet growing regions worldwide. Establishing uniform CLS disease pressure across a field trial is necessary to conduct reproducible field research for screening varieties as well as evaluating fungicide management programs. In 2022, a field trial was conducted to evaluate the use of grain sorghum for preparing *C. beticola* inoculum. A randomized complete block design with four replications was used, and a moderately susceptible sugarbeet variety with a CLS rating of 4.9 was sown on May 24. Four isolates of *C. beticola* were grown on autoclaved sorghum grains in breathable polypropylene bags with a filter. On July 13, treatments were applied with a modified duster over each row and included *C. beticola*-infested sorghum grains milled into a powder and mixed with fine talc (2:1 w/w) applied at 4.5 lbs/A and 9 lbs/A, and without fine talc applied at 4.5 lbs/A. In addition, *C. beticola*-infested whole sorghum grains were broadcast at 11 lbs/A. Treatments were compared to a non-inoculated control and a standard inoculation method of dried ground CLS-infected sugarbeet leaves mixed with fine talc (2:1 w/w) applied at a rate of 4.5 lbs/A. There were significant differences for disease severity, yield, and recoverable sucrose per acre. All treatments with *C. beticola*-infested sorghum grains resulted in higher disease levels than the CLS-infected sugarbeet leaves. The non-inoculated control resulted in the lowest level of disease. The use of sorghum grain allowed for greater control of inoculum quality and resulted in higher disease pressure than the CLS-infected leaf inoculum. Additionally, the broadcast application does not require mixing with talc and significantly reduces the time needed for inoculating large areas. This trial will be repeated in 2023.

MATSUHIRA, HIROAKI*, Hokkaido Agricultural Research Center, National Agriculture and Food Research Organization, 9-4 Shinseiminami Memuro, Hokkaido 082-0081, Japan.

Marker-assisted selection of O-type candidates resistant to yellowing disease by Beet leaf yellowing virus in Japan.

Yellowing disease, caused by viral infection, is a problem in sugar beet production. In Japan, yellowing disease caused by infection with Beet leaf yellow virus (BLYV), similar to Beet western yellows virus, occur periodically, and result in low yields. Therefore, development of varieties resistant to yellowing disease is desired. Previously, we found QTLs linked to yellowing disease resistance using a genetic resource of USDA. In this study, we tested the effectiveness of molecular markers linked to the QTLs for selection of yellowing resistant breeding line. We crossed our O-type (maintainer of Owen CMS) lines susceptible to yellowing disease, with resistant genetic resources of USDA to create segregating populations. Their resistance to yellowing were tested by inoculation test of BLYV via peach aphids. The yellowing index (YI) of susceptible O-type were above "2.5" (0: no disease symptom to 3: 100% yellowing of mature leaves), whereas the resistant genetic resource had YI less than "1.0". The F_3 lines showed variation in YI ranging from "0.3" to "2.9". Genotyping of markers linked to resistance revealed that the F_3 lines having resistant genotype of the markers averaged YI of "0.7", whereas F_3 lines having susceptible genotype of the markers averaged YI of "2.0". These results demonstrated that the markers can effectively select yellowing resistance. These lines were further selected for sugar content, restorer-of-fertility genes, and monogerminity to obtain O-type candidates. The agronomic characteristics of these lines will be determined in the future. This research was supported by the research program on development of innovative technology grants from the Project of the Bio-oriented Technology Research Advancement Institution (BRAIN).

METZ, NICHOLAS*[1], AVNEESH KUMAR[2], OLIVIA TODD[3], BRADLEY SOWDER[2], ANN FENWICK[2], AMY NIELSON[2] and KEVIN DORN[4], [1]USDA-ARS, 1701 Centre Ave, Fort Collins, CO 80526, [2]USDA-ARS, 1701 Centre Ave, Fort Collins, CO 80526, [3]UDSA-ARS, 1701 Centre Ave, Fort Collins, CO 80526, [4]USDA-ARS, 1701 Centre Ave, Fort Collins, CO 80526.

Genomics-driven mapping of *Rhizoctonia solani* resistance in sugar beet.

Sugar beets are attacked by several fungal pathogens that cause root damage, including *Rhizoctonia solani*, which causes the disease Rhizoctonia Crown and Root Rot (RCRR). Several Quantitative Trait Loci have been reported to harbor resistance genes to RCRR, but no RCRR resistance gene has been cloned from sugar beet or a crop wild relative. Discovering the identity of the genetic elements responsible for resistance will enable more precise molecular assisted breeding alongside a better understanding of the plant-pathogen interaction to drive future RCRR research efforts. In this study, we report a multifaceted genomics approach to map and characterize RCRR resistance from a highly resistant sugar beet prebreeding line. Utilizing bi-parental mapping, an RCRR susceptible parent line (03-124) was crossed with the highly resistant FC709-2 to produce F2 families. Multiple F2 families have been subjected to both field and greenhouse disease trials to study phenotypic segregation. Multiple genomic strategies, including Quantitative Trait Loci mapping using genotyping-by-sequencing and bulk segregant sequencing as well as genotyped using genotype by sequencing, were used to map the genomic loci harboring RCRR resistance in this population. To avoid reference genome bias, as well as provide a new tool for sugar beet pan-genomics and trait mapping, we developed the most complete and contiguous genome assembly for sugar beet to date from FC709-3, a more inbred and homogeneous derivative of FC709-2. Finally, we performed a comprehensive RNA expression profiling experiment using Illumina RNA sequencing to provide transcriptional support for genome annotation and to identify differentially expressed genes that are co-located in RCRR QTL in this mapping population. We anticipate that an enhanced understanding of the genetic sources of RCRR resistance will drive marker assisted selection in both pre-breeding and commercial breeding programs, as well as drive the discovery of novel RCRR resistance traits from crop wild relatives.

METZ, NICHOLAS*[1], KEVIN DORN[2], ANN FENWICK[1], BRADLEY SOWDER[1], AMY NIELSON[1], OLIVIA TODD[1] and AVNEESH KUMAR[1], [1]USDA-ARS, 1701 Centre Ave, Fort Collins, CO 80526, [2]USDA-ARS, 1701 Centre Ave, Fort Collins, CO 80526.

Fine tuning artificial Rhizoctonia inoculation methods to improve disease pressure using DNA-based quantification.

Rhizoctonia solani is a major pathogen plaguing sugar beet production. This pathogen causes crown and root rot leading to plant death and reduced yields. Many breeding efforts across the world use artificially inoculated disease nurseries to evaluate germplasm for resistance. These fields are typically in rotation with other crops and inoculated at a set rate to induce disease. The general assumption is that there is little to no *Rhizoctonia* in the soil, but by measuring the baseline pressure, we can accurately apply inoculum at an optimized rate. Understanding baseline disease pressure is critical for adjusting inoculum to achieve uniform disease pressure from year to year. In this study, we conducted an inoculum dosage curve experiment using commercial check varieties with six different application rates. This results from this trial will guide the optimization of inoculum application rate to ensure even uniform disease pressure across years. Soil samples were taken from this experiment, as well as from the site of the recurring USDA-ARS Fort Collins *Rhizoctonia solani* nursery, to quantify pre- and post-inoculation pathogen load. Using qPCR we can distinguish the levels of Rhizoctonia in the soil for different stages within our testing year as well as our crop rotation. Gaining information on the baseline levels of disease within our nurseries allows for more accurate selection and evaluation of germplasm for Rhizoctonia Crown and Root Rot resistance.

NAEGELE, RACHEL P.* and MICHAEL PETERS, USDA-ARS, 1066 Bogue Street, #360, East Lansing, MI 48824.

Phenotypic predictors for breeding drought tolerance in juvenile sugar beet.

Prior to establishment, juvenile sugar beets are highly sensitive to environmental stress such as drought. We evaluated populations of six breeding lines for tolerance to drought stress in the greenhouse over 3 weeks. Plants were phenotyped for photosystem I and II activity, leaf temperature and humidity, greenness, plant height, wilting, and root weight for the duration of the experiment to identify predictors associated with improved stress response. Differences were observed within and among beet lines with most beets showing a significant reduction in photosystem activity, GSW, and root weight during drought stress, and a significant increase in wilting. However, the extent of the stress response varied greatly among and within lines.

NARIHIRO TSUBASA*[1], HIROAKI MATSUHIRA[1], SANO MASAKAZU[1], KITAZAKI KAZUYOSHI[2], KUBO TOMOHIKO[2] and KURODA YOSUKE[1], [1]National Agriculture and Food Research Organization, 3-1-1 Kannondai, Tsukuba, Ibaraki 305-8517, Japan, [2]Hokkaido University Kita 8, Nishi 5, Kita-ku, Sapporo, Hokkaido, 060-0808 Japan.

Introduction of BLYV resistance to 3-way crossed F1 generation by resistant pollen parent.

Beet leaf yellowing virus (BLYV) causes yellowing of sugar beet leaves and decrease about 30% of sugar yield in severe case. The virus incubates in weeds and green peach aphid (GPA) spread the virus to sugar beet. It is difficult that extermination of the virus and the vector, so utilization of the resistance to BLYV is attracting attention. It is known that *Beet western yellows virus* (BWYV) resistant genetic resources which were published by USDA also have resistant to BLYV. Our group doing breeding of pollen parents and seed parents using the genetic resources.

Sugar beet cultivar is generally 3-way crossed F1 generation, so their half of genetic background is derived from pollen parent. If the resistance has dominant effect, we can easily introduce BLYV resistance to cultivar from pollen parent. In this research, we report the resistance test result of 3-way crossed F1 materials whose pollen parent is BLYV resistant. In 2021, we tested 6 standard materials and 16 3-way crossed F1 materials which were derived from 2 BLYV resistant pollen parents. In 2022, we tested 6 standard materials and 17 3-way crossed F1 materials which were derived from 7 BLYV resistant pollen parents. The virus was inoculated to sugar beet seedlings by GPA. The seedlings transplanted in the field. The field design consisted of 4 replication whose plant number was 5. In 2021, non-inoculated field also made as same design. In 2021, we evaluated yellowing symptom of inoculated plants and compared yield between inoculated plots and non-inoculated plots. In 2022, we evaluated yellowing symptom of inoculated plants. For evaluation of yellowing symptom, we defined 10-level Yellowing Index (YI). YI 0.0 means no-symptom, YI 3.0 means the severest symptom. For evaluation of decrease of yield, the weight of root and crown, Brix were measured. Yellowing evaluation was done at 2021/8/31, 2022/8/17, harvest day was 2021/10/25. In 2021, YI, reduction rate of root and crown weight and Brix of the "Strong" resistant standard lines were 0.0-0.1, 6.2-6.6%, 1.1-5.3%, those of the "Medium" resistant standard lines were 0.0-0.1, 6.2-6.6%, 1.1-5.3%, those of the "Weak" resistant standard lines were 2.1-2.7, 43.7-58.9%, 13.4-13.9%, those of 3-way crossed lines were 0.1-0.6. 20.5-36.9%, 5.8-9.8%. In 2022, YI of the "Strong" resistant standard lines were 0.4-0.2, those of "Medium" resistant standard lines were 1.1, those of "Weak" resistant standard lines were 1.9-2.2, those of 3-way crossed lines were 0.4-0.9. From this research, using BLYV resistant pollen parent, introduce of resistance to 3-way crossed generation and reduction of yield decrease were succeeded. The resistance shows dominant effect, so introduction of BLYV resistance by resistant pollen parent can be promising methods. This research was supported by the research program on development of innovative technology grants from the Project of the Bio-oriented Technology Research Advancement Institution (BRAIN).

NATWICK, MARI B.*[1], GARY A. SECOR[1] and MELVIN D. BOLTON[2], [1]North Dakota State University, Department of Plant Pathology, NDSU Dept 7660, PO Box 6050, Fargo, ND 58108, [2]USDA-ARS, 1616 Albrecht Blvd. N, Fargo, ND 58102-2765.

Agrobacterium-mediated expression of necrotic *Cercospora beticola* proteins.

The most economically important disease of sugar beet is Cercospora leaf spot (CLS) caused by the hemibiotrophic fungal pathogen *Cercospora beticola*. The pathogen causes foliar damage via necrotic lesions that, if left unmanaged, can significantly reduce crops yields. The lesions are caused by a plethora of effectors molecules, some of which have been previously characterized. The process of characterizing effector proteins can be tedious and involve difficult extraction and protein purification processes. In order to streamline the process of identifying putative effectors and their functions, transient gene expression methodologies can be employed. One such technique relies on the natural ability of *Agrobacterium tumefaciens* to transfer DNA into plants that will be subsequently integrated into their genomes. This ability has been coopted by researchers to express genes of interest in plant cells. Here, we developed lines of *A. tumefaciens* that express candidate *C. beticola* effector genes in the model plant *Nicotiana benthamiana*. When plants express a necrosis-inducing protein, cell death will occur in the plant tissue at the zone of infiltration. Consequently, this strategy allows for a high throughput means to screen *C. beticola* candidate effector proteins for the ability to induce necrosis. This strategy and the identification of *C. beticola* necrosis-inducing effector genes will be discussed.

O'BOYLE, PATRICK* and MARGARET REKOSKE, KWS Seeds, LLC., 1325 Valley View Road, Shakopee, MN 55379.

Gaining ground against Cercospora: A decade of breeding for Cercospora Leaf Spot tolerance.

Cercospora leaf spot (CLS) of sugar beet, caused by *Cercospora beticola*, is one of the most devastating diseases of sugar beet worldwide. This polycyclic fungal disease is controlled via a combination of cultural, chemical, and genetic means. Typically, genetic tolerance, while a critical part of an overall disease control program, has been limited due to an association between high disease tolerance with low yield performance under non-disease conditions. Progress has been made over time, and in recent years high performing hybrids with strong tolerance have been available that provided an increased level of CLS tolerance without sacrificing performance in the absence of CLS pressure. Recently, the introduction of CR+®™ hybrids has provided a new valuable tool against Cercospora infection, offering significantly enhanced levels of CLS tolerance combined with even higher performance regardless of disease pressure. Hybrids grown commercially in Michigan over the last decade were compared head-to-head in yield and disease nursery trials illustrating the breeding gains over time and higher relative performance in strong CLS hybrids. More recently, initial CR+®™ hybrids were competitive with traditional strong CLS tolerant hybrids for sugar yield and slightly weaker in sugar content, with a significant reduction in relative CLS rating (in % of benchmarks). Subsequent CR+®™ OVT entries pending commercial approval have shown a statistically significantly higher sugar content (LSD 0.05%) compared to initial CR+®™ offerings, and an increase of over 10% (relative to benchmarks) in sugar yield, while staying under 50% of the benchmark level for CLS rating.

STRAUSBAUGH, CARL A.*[1] and ERIK J. WENNINGER[2], [1]USDA-ARS NWISRL, 3793 N. 3600 E., Kimberly, ID 83341, [2]Univ. of Idaho, Kimberly Research and Extension Center, 3806 N. 3600 E., Kimberly, ID 83341.

Beet curly top virus strains in sugar beets, dry beans, and beet leafhoppers along with vector population dynamics in southern Idaho.

At the request of a sugar beet industry stakeholder, beet leafhopper

(BLH) populations in southern Idaho were tracked in four counties during the 2020 and 2021 growing seasons in desert areas and sugar beet and dry bean fields in southern Idaho with yellow sticky cards. Samples were collected on a weekly basis from mid-April through mid-September to assess all leafhoppers for population levels and the presence of Beet curly top virus (BCTV) strains. Crop plants from monitored fields were also assessed for the presence of BCTV strains. Once BLH populations in Elmore Co. began increasing in May, they were present in at least double-digit numbers through most of the summer at all sites both years. However, the BLH numbers at desert sites in other areas were at or near zero. In areas with low BLH desert populations, local weed populations appeared to be the primary source of BLH in crop fields. Preliminary data suggest two haplotypes (based on cytochrome oxidase gene) dominate the BLH population. Over the 22-week collection period, horizontally oriented cards averaged 75% and 51% fewer BLH than vertically oriented cards in 2020 and 2021, respectively. In 2020, 42% of the BLH samples were positive for the BCTV coat protein and Worland was the dominant strain. The phytoplasma, morphotyping, and 2021 BCTV strain identifications are currently a work in progress. Once all data are collected, the project will establish the BCTV strains for which host plant resistance is needed and the time when sugar beets are at highest risk for BCTV infection.

VINCILL, ERIC D.* and LAURIE K. JACKSON, USDA-ARS, NWISR, Kimberly, Idaho 83341 USA.

The sugar beet germplasm line EL10 is amendable to agroinfiltration-mediated transient expression with a capacity to achieve high levels of transgene expression in the leaf.

Since discovering *Agrobacterium tumefaciens* has the distinctive capacity to incorporate a specified part of their transfer-DNA (T-DNA) into the genome of plants, the bacteria has been used extensively to genetically transform into crops agronomically important traits such as disease resistance, herbicide tolerance, and increased nutrition and yield. While Agrobacterium mediated plant transformation has proven its worth to crop enhancement and breeding platforms, some crop species such as sugar beet are recalcitrant to agrobacterium mediated transformation or transformation events are so infrequent that it makes it cost/time prohibitive for most plant biology labs to pursue. Therefore, determining the factors influencing efficient agrobacterium transformation in sugar beet would prove useful in the establishment of sugar beet transformation pipelines. Here, we report on the identification of a sugar beet germplasm 'EL10' (USDA-ARS PI 628755) that is amendable to agroinfiltration-mediated transient expression and is capable of achieving high levels of transgene expression in the leaf. Factors influencing efficient agroinfiltration were investigated including different strains of *Agrobacterium*, media composition, bacteria/plant co-incubation and plant growth conditions. Furthermore, by incorporating optimal promoter/5′-leader and terminator elements into the transgene T-DNA cassette, a significant reduction in transgene silencing was observed allowing for high expression of transgene to be maintained for as long as 5 weeks after agroinfiltration. The significance of 'EL10' germplasm (and not other sugar beet germplasm lines investigated) having the ability to be transiently transformed by *Agrobacterium tumefaciens* will also be discussed. Overall, the results presented will aid in the establishment of a routine agroinfiltration-mediated transient method for sugar beet. ☐

ABSTRACTS: PROCESSING

ORAL PRESENTATIONS

ANTENS, NICO*[1] and GARSIA, CHRISTIAN[2], [1]Cosun Beet Company, [2]Royal Cosun, [1]Noordzeedijk 113, 4671 TL Dinteloord, The Netherlands, [2]Kreekweg 1, 4671 VA Dinteloord, The Netherlands.

Practical challenges of using vision systems for new applications in a sugar factory.

Vision systems, or so-called image processing or image analysis systems, are very common in the industry, most often to control product quality. Within the sugar industry the amount of commercially available applications are however still rather limited. Of course, on the sugar end the image processing systems for sugar color and crystal size characterization are well developed and integrated in the existing production process. Further, in the past for example a vision system was developed to determine the top tare of the beet in the tare house to skip manual topping. Since the introduction of the smartphone and the related apps, image processing is very common and makes part of our normal live. As a result, new ideas for vision system solutions and applications within the sugar factory popped-up by employees in recent years. Three possible applications were further investigated and developed in the Dutch factories of Cosun Beet Company: beet yard volume estimation, foreign object detection between unwashed beets and the size measurement of beet cossettes. This paper describes the problem/solution, the method and results of the performed plant trials and the related practical challenges. It is certain that a joint approach from the sugar industry is required to develop commercially available applications.

ARRIETA, IRMA*, PETER FERRERO and CY GAUDET, Amalgamated Research LLC, 2531 Orchard Drive East, Twin Falls, ID 83301, USA.

Reduced capital and operating costs for beet molasses desugarization: Pilot scale optimization.

The desugarization of beet molasses through simulated moving bed (SMB) technology has been practiced commercially in the United States for more than thirty years. Currently, most U.S. beet sugar producers operate some form of molasses desugarization system. Substantial improvements in the capacity and performance of these systems have been made over the years. However, this development process has typically been based on an expert-guided approach to optimization, which is not a tactic that is necessarily guaranteed to find an optimal set of operating conditions. Recent research and development efforts at Amalgamated Research have instead taken a rigorous and comprehensive approach to the optimization of a coupled loop beet molasses separator system. This experimental methodology, combined with in-house mathematical modeling capability, has led to significant advances being made to reduce both the capital and operating cost of molasses desugarization systems. Some of these improvements have been confirmed by testing on a full industrial scale with an existing second loop of a molasses separator. Further loop 1 factory trials are planned for the 2022-23 campaign.

ATKINSON, MARQUIS G.*[1] and SCHOENFELDER, CARL J.[2], [1]The Amalgamated Sugar Company, 50 South 500 West, Paul, ID 83347, [2]Hydrite Chemical Company, 17385 Golf Parkway Brookfield, WI 53045.

Utilization and opportunities of oxidizing biocide for process microbial control.

Microbial pressures from beets entering the process through the cossette mixer and/or diffuser are a major contributor to reduction of recoverable sugar, in both unaccountable loss and sugar loss to molasses. Even with proper temperature profiles and process management parameters, infections can warrant the need for chemical addition to effectively address. Many current chemical approaches rely on products that are technically only biostats, have limited species efficacy, are prone to resistance build-up, or even become feedstock to the microbe itself. Additionally, those products that are true biocides have historically carried issues with usage or handling limitations, related to undesirable downstream bi-products or contribution. An alternative approach that addresses the above-mentioned concerns and product limitations is to use an oxidizing acid (peracetic acid) as a replacement for conventional and historic programs. Taking advantage of the chemistry's favorable properties and applying them to a well-structured microbial control program, results can be obtained that are competitive or superior to conventional programs, in terms of efficacy and cost. An extended trial was conducted at the Owner's facility to validate the true process impact of this chemistry, as well as identifying the infrastructure requirements, injection locations, and other ancillary considerations. The results of this trial confirmed the chemistry's viability potential and identified other valuable considerations for future product utilization on a sustainable basis.

BAKER, MAX* and IQBAL, SHAHRUM, Bray International, 13333 Westland E Blvd, Houston, Texas 77041.

Accurate milk of lime dosing - A study into valve solutions for milk of lime control.

Milk of lime is used to capture and remove impurities in the juice of sugar cane. Accurate lime dosing is essential to ensure an efficient, high quality sugar juice purification stage and a pH value of between 7.5-8.2 for the mixed juice. The properties of milk of lime cause scaling, jamming and erosion of flow control equipment like pipes, valves and pumps. More often than not, this leads to an increased cost of maintenance, loss of milk of lime, and inaccurate level of pH of the mixed juice affecting the equipment and extraction of sugar crystals. In this presentation, I will review the challenges which valves face due to scaling, media build up, and internal wear and tear from the suspended solids in milk of lime. Additionally, I will review valve solutions which are able to mitigate these issues while enhancing the service life and reliability of the valve. These solutions revolve around the inclusion of specialty materials and design modifications within Bray's highly engineered products created for use in the Sugar industry.

BUSCHETTE, LYNN*, DAVID R. GROOM, Beverly Jacobson, and Stephanie Heiden. American Crystal Sugar Company, 1700 North 11th Street, Moorhead, MN 56560.

A field to factory study of the microbial flora of sugar beets and factory processing streams utilizing MinION deep sequencing technology.

American Crystal Sugar Company (ACSC) factories have experienced periods of high unaccountable sucrose losses with consequent lower than expected recovery. In addition, physical processing issues have been intermittently encountered such as poor syrup filtration that have no readily identifiable causes. Increased unaccountable losses have been correlated to increases in volatile fatty acids such as lactic acid, invert sugars, and ethanol in the diffusion juice and subsequent process streams. Elevated sucrose losses tend to occur when cossette ethanol levels increase along with invert sugars. The prior described events may or may not present as physical operating issues. The response to various treatments to manage undesirable microbial activity during the process can vary greatly. In the spring of 2019 PCR work was undertaken to identify microorganisms present in the beets, the soil, and the process streams, with the aim to correlate the presence and abundance of bacterial taxa with beet health and processing issues. Various classes of microorganisms were identified in both the beets and process streams that included anaerobic and aerobic mesophiles and thermophiles along with aerobic psychrophiles. 98 species were identified that included *Leuconostoc, Bacillus, Enterococcus, Pseudomonas, Thermoanaerobacter* spp. In the fall of 2020 work was done with USDA in utilizing deep sequencing of 16S rRNA gene PCR amplicons with the MinION sequencer (Oxford Nanopore Technologies). Results of the work were presented at the 2021 ASSBT conference. Work has continued at ACSC utilizing MinION technology. In 2021 samples from storage piles (from brei) were analyzed in addition to factory processing samples. *Lueconsostoc, Microbacterium, Pantoea, Pseudomonas* spp. were identified and quantified. To date sequencing experiments have demonstrated that there is a great deal of variability in the micro flora found in field and process streams, the populations shift across campaign.

BUSCHETTE, LYNN*, DAVID R. GROOM, BEVERLY JACOBSON, LAURA HAGAN, AND SETPHANIE HEIDEN. American Crystal Sugar Company, 1700 North 11th Street, Moorhead, MN 56560.

Storage and process operating condition impact on sugar beet microbial flora metabolites.

American Crystal Sugar Company (ACSC) factories have experienced periods of high unaccountable sucrose losses with consequent lower than expected recovery along with physical processing issues such as poor syrup filtration. Work is ongoing to identify the organisms involved and identifying metabolites associated with organisms identified. Analytical chemistry is more easily and rapidly obtained. In our studies we attempt to link specific organisms with various metabolites produced at a given set of processing conditions. We have found that under varying operating conditions a given organism can produce different metabolites that in turn impact the process differently. For instance, *Leuconostoc* may produce slime that interferes with filtration and settling under one set of conditions in another scenario a metabolite such as ethanol is produced with no physical manifestations, but an unaccountable loss is noted. Increased unaccountable losses have been correlated to increases in volatile fatty acids such as lactic acid, invert sugars, and ethanol in the diffusion juice and subsequent process streams. Elevated sucrose losses tend to occur when cossette ethanol levels increase along with invert sugars. The prior described events may or may not present as physical operating issues. The response to various treatments to manage undesirable microbial activity during the process can vary greatly. The essence of the studies completed is to define the relationships between microbes, environment, and metabolites produced.

CHUDASAMA, ARVIND*, S&P Global, Ropemaker Place, 25 Ropemaker St, London, SE1 0SU, UK.

Sugar reduction/replacement technologies seek to mimic a single ingredient with a concoction.

Demonization of sugar has given birth to the sugar reduction sector, and in the recent past, accelerated it. Over the past 10 years, there has been a significant rise in new products claiming to be 'low calorie', 'low sugar', 'no added sugar', 'sugar free'. In 2018, the market researcher Innova Market Insights found that in the US, 8% percent of all new food and beverage launches featured a sugar reduction claim. Of these, 42% of the products claimed to have "no added sugar", while 36% were "sugar-free" and 27% "low-sugar". But it is unlikely that the sugar reduction sector will expand much beyond the current level. This is largely because of the technological, regulatory, economic and consumer acceptance hurdles a new ingredient has to traverse. Sugar reduction strategies include removing or reducing the amount of added sugar, replacing part of the sugar formulation with non-nutritive sweeteners or using novel processing technologies. Of the several hurdles, consumer acceptability is critical. To offset the evident lack of palatability of intense sweeteners (which often carry metallic or bitter taste and aftertastes) in sugar-reduced products, sweetness modulators are used. How long before consumers wise up to the fact that sugar reduced fare comprises a concoction of ingredients to replicate the taste of sugar? This presentation exposes the lengths to which companies are going to mimic the taste and function of sugar in various products.

DEROSIA, TARYN* and TONY WIDBOOM. Barr Engineering Co., 170 South Main St, Suite 500, Salt Lake City, UT 84101.

Greenhouse gas accounting considerations for the sugarbeet industry.

Greenhouse gas (GHG) emissions and carbon footprints are a growing area of focus for many economic sectors in the United States. The carbon footprint of products may influence how environmentally conscious consumers spend and lenders invest their money. As a result, many industries are pivoting their messaging to incorporate environmental initiatives involving reductions of GHG emissions. Food product manufacturers using beet sugar are requesting GHG information from their supply chain to define and improve their carbon footprints. Developing a carbon footprint through GHG accounting is not a standardized procedure and will vary between industries and from facility to facility depending on the activities that occur. However, sugar beet processing facilities across the country have some common practices and activities that can be considered when developing a carbon footprint. The sugarbeet industry is also unique from other industries due to their common structure as a cooperative of growers, including both agricultural operations and processing. GHG accounting is separated into scope 1, 2, and 3 emissions. Scope 1 emissions are those directly from the operations. Common scope 1 emissions in the sugarbeet industry include factory fuel combustion, vehicle emissions, agricultural practices such as fertilizer application, and carbon sequestration from crop growth. Scope 2 emissions are from third-party electricity suppliers. Scope 3 emissions include downstream and upstream sources such as GHGs from fertilizer manufacturing or transport to the end users. The methodology for developing a carbon footprint for the sugarbeet industry is reviewed. Developing scope 1, 2, and 3 activities is the first step in developing a GHG accounting protocol for a facility. Site-specific emission factors and inputs are ideally used to develop a GHG inventory. Site-specific emission factors may be based on stack testing, material balances, or analytical results. Inputs may be based on accounting or other facility operational records. If site-specific data is not available, calculations may be completed based on available literature from sources such as the International Panel for Climate Change, Environmental Protection Agency guidance, or other studies and literature. As a facility develops and tracks its GHG inventory, the information may be used to provide carbon footprint information to consumers, develop marketing materials, and identify sources for GHG reductions.

SIGRID BAETH*[1], TRENT HOLCOMB[1], SCOTT HYER[1], MARK VANDERHOFF[1], BRIAN ALLENDER[1], TYSON SUMMERS[2] [1]Amalgamated Sugar, 138 West Karcher Road, Nampa, ID 83687, [2]Hydrite, 17385 Golf Parkway Brookfield, WI 53045.

Steam side boil outs of carbon fouled evaporators.

Evaporator cleanliness is crucial to the efficient operation of a sugar factory. Fouled evaporators can impact everything from beet end heat balance, pan boiling, byproduct drying, ect. Fouling of the juice sides of evaporators has been happening since sugar beets have been being processed. The composition of the scale on the juice side is commonly a form of calcium scale. The process for removing calcium scale is well established and proven, while the steam side is not something that is routinely done. Analysis of the steam side fouling showed >95% organics. This would require a different boil out chemistry than the proven solution for the metal scales. For this we decided to apply Hydri-Maize® CIP 3421, a peroxide-based solution, with caustic. When working with a peroxide solution safety needs to be well thought out. Peroxide will expand over 10x when applied to organic material and will be very exothermic. Having a place to go with the expanded material along with all the material that will be removed is very important. Hydri-Maize® CIP 3421 with caustic showed to be a very effective combination with evaporator efficiencies returning to design the following campaign.

HOWE*[1] and KOCHERGIN[2], [1]BetaVA Ltd, 16 Holmes Drive, Ketton, Stamford, [2]Amalgamated Sugar LLC, S Saturn Way, Suite 100, Boise, Idaho, 83616.

A path to implementation of the lime free sugar process.

In the late 1990's, an international consortium of nine sugar companies spent several years investigating and scalable piloting of a new lime-free sugar beet process, that was initially proposed by Amalgamated Research Inc. (ARi). The main feature of the process was the replacement of the lime-based juice purification by a chromatographic process. Filtered and softened raw juice was concentrated and separated using resin technology. As a result, non-sugar removal exceeded 80 %, which was far superior to the typical 30% non-sugar removal for conventional beet sugar operations. The chromatographic extract of 98 purity allowed three white sugar boiling's similar to that in a cane sugar refinery. The configuration of the lime free process was extensively studied and presented a sugar factory with much improved efficiency and different process configuration, although the required equipment was largely that currently seen within the industry. At the end of the eight-year study, the process was considered ready for commercial implementation but as of 2022 no business has yet had the vision or taken the step of realising the sustainable advantages it presents. Under increasing environmental pressures for the beet sugar industry and the challenge to reduce the carbon footprint it is believed the lime free sugar process may offer an attractive opportunity. This presentation reviews the current status and proposes a potential path to implementation. It will also review the challenges encountered and overcome in the original study work. A discussion will follow how the lime free sugar process may become a low risk future reality for sugar beet processors.

HYER, SCOTT*[1], SIGRID BAETH[1], IRMA ARRIETA[2] and CY GAUDET[2], [1]Amalgamated Sugar Company, 138 W Karcher Rd, Nampa, ID 83687 and [2]Amalgamated Research LLC, 2531 Orchard Dr E, Twin Falls, ID 83301.

Reduced capital and operating costs for beet molasses desugarization: Industrial results from an existing separator.

Traditionally, Amalgamated Research uses a coupled loop beet molasses desugarization system (simulated moving bed (SMB) chromatography system) to separate both betaine and sucrose from the feed molasses. In recent years, Amalgamated Research has carried out extensive pilot testing to rigorously optimize the design and operation of beet molasses desugarization systems. This has led to the development of a six-cell design for the second loop, which yields robust performance at increased capacity and at reduced water usage. This new six-cell configuration for the second loop has been implemented on an industrial scale at the Amalgamated Sugar factory located in Nampa, ID. Despite the Nampa system being a relatively old Amalgamated Research design, six-cell operation showed no reduction in performance compared to eight-cell operation, thereby confirming the robustness of the results obtained on a pilot scale. The water used for the separation was also reduced by 10% from the standard design value, without any negative impact on loop 2 performance. Operation in six-cell mode has now become standard operation at Nampa. The successful factory trial demonstrated that six-cell loop 2 operation can achieve a sucrose purity and recovery comparable to that of a conventional eight-cell separator, at an increased throughput and at reduced water consumption. This represents a 25% reduction in equipment size for the second loop and a significant reduction in both capital and operating cost compared to the previous eight-cell design.

LARSEN, KASPER GEHL, EnerDry A/S, Kongevejen 157, 2830 Virum, Denmark

How to reduce CO2 emissions from sugar factories, to meet clients and authority's requirements.

In the years to come there will be a political requirement for the industry to emit less CO2, and the political goal is CO2-neutral production, sooner or later. Sugar factories that ignores the CO2 emissions, risk to be out of business in the long run. Carbon labelling on consumer products, might result in diversified prices, depending on the CO2 emissions for the specific sugar factory. The first step in becoming CO2 neutral is to replace the existing drum dryer with a steam dryer. The cost of this is app. 250 € for the yearly saving of a ton of CO2. At this first step the CO2 has been reduced by app. 33% from the factory. The steam dryer is proven technology that has been in industrial operation since the 90´ties. Second step is to burn the steam dried sugar beet pulp in the boiler, and the factory will be CO2 neutral, except for the limekiln. There will be some excess sugar beet pulp, depending on the sugar factory efficiency. Biogas is an alternative that can be considered. If all pulp is made to biogas, app. 70% of the fuel required can be produced. Fossil fuel, or biogas from other product must be added. With no pulp for sale to the feed marked, this feed must be produced elsewhere. This leads to an increase in CO2 emissions, as agriculture will pollute while producing this missing feed, therefore no net CO2 reduction. That is the feed or fuel discussion. As a future perspective, the sugar factories with the gasifier, can in the offseason produce a gas that can be utilized in a methanol reactor. This power-to-X is not relevant yet, but might become big business in the future. The fuel in the offseason could be straw, wood chips or other CO2 neutral fuel.

LEHNBERGER, ANDREAS[1*] and SCHMIDT, HANS[2], [1]BMA Braunschweigische Maschinenbauanstalt AG, 38122 Braunschweig, Germany and [2]BMA America Inc., Aurora, IL 60504, USA.

Improved energy efficiency in the sugar house.

The energy input in the sugar house has a crucial influence on the energy demand for sugar production in the beet sugar factory. In view of the changing political and economic boundary conditions, concepts to increase the energy efficiency in the sugar house are being discussed. It suggests itself here to minimize the quantity of water to be evaporated in the sugar house area. There are also interesting concepts featuring continuous crystallization and mechanical vapor compression. Such considerations are geared to increasingly utilize renewable electric energy instead of thermal energy which today mainly comes from fossil fuels. In times of increasing energy costs, a low specific energy demand represents an economic advantage for beet sugar factories. The politically induced reduction of the use of fossil fuels also for sugar production boosts the discussion of feasible and necessary adaptations in the sugar production process. This paper focuses on the energy demand of the sugar house, which bleeds the largest part of vapor from the evaporators. Conventional and new concepts to minimize the energy demand required for crystallization and the separation of crystals and syrup will be presented. They will be discussed in the light of the new boundary conditions. Decisive quantities for the steam demand include the water intake from thick juice and the demand for melting B and C sugar, as well as the quantity of wash water used at the centrifugal machines. For several decades, various crystallization concepts have been applied to achieve an automatized and to a large extent reproducible crystallization process. Depending on the desired crystal size, one-stage or two-stage crystal seeding processes are often applied, which primarily serve to ensure a feed solution with maximum dry substance content. In this way, the process design allows for minimizing the water intake and thus the steam demand. The application of continuous crystallization plays a crucial role, especially in concepts that include mechanical vapor compressors. Here, the thermal energy supply (steam) to the sugar house can be reduced by means of electric energy usage.

MARSH, DANIEL L.*[1], SMITH, JEREMY D.[2], HILLMAN, KYLE M.[1], THORNE, KEVIN M.[1], [1]ChemTreat, Inc. 5640 Cox Road, Glen Allen, VA 23060 and [2]Amalgamated Sugar Company, 50 South 500 West, Paul, ID 83347.

On-line process contaminant monitoring and automation in beet sugar factory steam systems.

Amalgamated Sugar's Mini-Cassia factory in Paul, ID was experiencing periodic boiler feedwater contamination events between December 9 and 18, 2020, resulting in low-pH excursions to the boilers. Existing contamination event monitoring and identification methods historically resulted in slow reaction times by factory boiler operators, as well as subsequent boiler contamination. Once introduced to the boilers, organic contaminants break down into organic acids that quickly consume boiler water alkalinity and often depress boiler water pH to acidic conditions. At a depressed pH, iron is solubilized and potentially re-deposited on critical heat transfer areas of boiler internal surfaces, reducing efficiency and equipment longevity. In partnership with Mini-Cassia's water treatment solutions provider, a trial unit of ChemTreat's RADAR™ was first deployed during fall of crop year 2020. RADAR is ChemTreat's fluorescence-based sugar detection technology. Its ability to detect contamination events in drips (return condensate) ten minutes faster than the incumbent sugar detector instilled a high level of confidence in RADAR's detection capability among factory personnel. During the initial contamination events in December 2020 and subsequent boiler shutdowns, the high level of confidence in RADAR's capabilities led the factory to directly wire RADAR's signal output to the factory DCS with data and alarms also flowing through ChemTreat's CTVista®+ intelligent water management platform. Permanent RADAR sugar detectors were installed in critical locations during crop year 2021 to monitor drips immediately downstream of evaporators and upstream of drip return pumps during campaign and juice run. In addition to fluorescence, the permanent RADAR units monitor pH, conductivity, and ORP. Full auto-divert capability of drips during contamination events was implemented for crop year 2022. RADAR's ability to quickly detect contamination in December 2020 provided early warnings that allowed factory personnel to divert drips, use alternative boiler makeup supply, and employ emergency chemistry control procedures to keep the factory operational without experiencing a temporary shutdown caused by steam loss. Had RADAR been more-fully integrated before the contamination events in December 2020, the technology could have saved the Mini-Cassia factory an approximate 48 hours of downtime.

NIELSEN, BJARNE CHRISTIAN*, Neltec Denmark, Vestergade 35, 6500 Vojens, Denmark.

Dynamic optimization of batch centrifugal separation by purging monitoring.

Correct timing of the steps in a batch centrifugal cycle is important to get the best separation of crystals and mother liquor. Too early application of wash water will mix the wash water with the mother liquor, reducing the washing efficiency of the water. Too late application of the wash water will cause the film of mother liquor on the crystals to be dried out by air pressed through the crystal wall by the high G-factor. Too early or too late shift between low and high purity run-off will send too much non-sucrose back in the process or too much sucrose to molasses. Too early stop of spinning will leave too much humidity with the crystals, requiring more energy for drying. Too late stop of spinning will make it difficult to discharge the crystals. For factories with constant massecuite quality and constant mixer filling the centrifugal cycle can be determined once and be valid forever. For all other factories, static settings for the centrifugal cycle will lead to less perfect separation. Results from purging sensors applied on various massecuites are presented showing different purging patterns.

ROBERT ZIMMERMAN*[1] and DAN COOPER and UDITHA SENARATNE [2], [1]The Western Sugar Cooperative, 1221 8th Ave, Suite E, Greeley, CO 80631, [2]Veolia Water Technologies, 120 S. State College Blvd, Suite 174 Brea CA 92821.

Fort Morgan wastewater treatment system improvements.

The Western Sugar Fort Morgan plant had a very simple wastewater treatment system consisting of mud presses to remove solids from the beet washing loop along with unlined finger ponds with some aeration in them to treat the high BOD water. These ponds were located close to the South Platte River and in-situ treatment in the ground water showed significant treatment was occurring, total treatment could not be substantially proven. The State of Colorado has been working with Western Sugar to upgrade this treatment system to ensure the water quality reaching the South Platte River meets their requirements. As part of this, Western Sugar partnered with Veolia to provide the 1st stage of treatment using a CSTR Anaerobic digester with associated equipment. This was installed in 2020-21 and was started in early 2022. This paper will provide details and results from this installation as well as plans for the 2nd stage of treatment of aerobic treatment in lined ponds and the challenges Western Sugar still faces for the final treatment of the wastewater for discharge.

SHUEY, MICHAEL*, Amalgamated Sugar, 1951 S. Saturn Way, Suite 100

Elements of an industrial safety program.

This presentation will outline an effective industrial safety program specific to accident prevention. Elements within this safety program follow the guidance of OSHA's Process Safety Management regulation. These include operating procedures, training, incident investigation, process hazard analysis, mechanical integrity, management of change, process safety information, prestart-up safety reviews, contractor safety, emergency planning and response, employee participation, hot work, and compliance audits. This presentation will share examples of the activities within each of these elements.

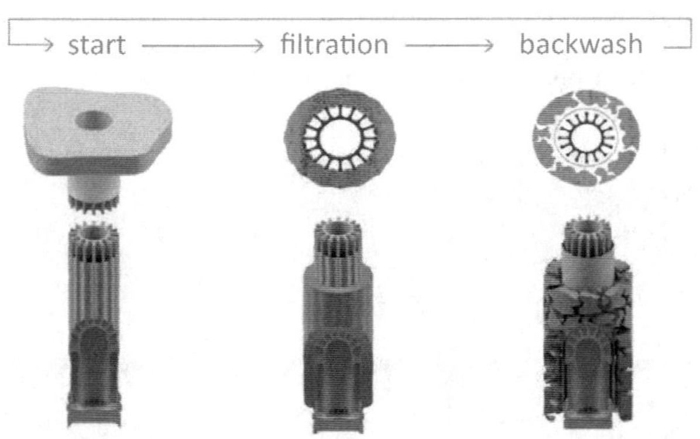

Patented candle design for unrivalled backwash efficiency | Filtration from outside to the inside | Backwash/cake discharge from inside to the outside

STEFAN SCHÖPF*, Lenzing Filtration, Werkstraße 2 4860 Lenzing Austria.

Cake & precoat filtration - Evaluating influences of increased backwash efficiency on cloth lifetime.

Raw juice, thin juice, thick juice, standard liquor and molasses are examples of liquid streams for which state of the art candle type cake/precoat filters are currently being considered as a technology upgrade to pressure leaf filters by sugar mills, particularly beet sugar ones. Despite the fully automated design of candle filters, especially in cake building applications such as 1st carbonation, cloth lifetime of the candle filter systems that have been on the market for decades has typically been extremely short. It is not uncommon for plants operating these older candle filter system designs to require cloth changes as often as thirteen times per campaign (120 days). Significant labor requirements as well as cost demands for replacement cloths have been experienced by sugar producers and processors using old design candle filters. Manufacturers of improved/upgraded designs of candle filters have allocated significant time and money resources to validating and optimizing backwash efficiency. In the chemical industry, for example, it has long been believed that design improvements in media backwash efficiency would directly contribute to improved filter performance. This hypothesis was recently validated on a pilot plant scale when a 3 w% $CaCo_3$ slurry was used to in a candle backwashing pilot test aimed at comparing the total weight of retained particles trapped in filter cloths following the backwashing of filter candles of three different designs. In this specific candle backwashing test, a sugar industry known candle design, one without a dip tube, was pilot tested. Another candle design commonly found in the chemical industry having a built-in dip tube was also tested. Thirdly, a new "channeled" patent-pending candle design also having a built-in dip tube was also tested. The outcome of the pilot test illustrated that the weight of the residual $CaCO_3$ particles per m^2 of filter cloth installed on the candle without a dip tube was 96 mg/m^2. That of the candle with dip tube commonly used in the chemical industry was 67 mg/m^2. The new "channeled" candle having a dip tube was only 22 mg/m^2. While these the total particle weights are interesting in absolute terms, especially interesting is the ratio of the total residual particle weights of the different candle designs since better backwash efficiency equates to longer filter cloth lifetime. In addition to closer look on above mentioned trials, the results of a five-month trial (September 2022 to January 2023) during the 2022/2023 campaign at a sugar factory in Germany will be summarized at the upcoming ASSBT conference in Savannah.

STOKES, ISAAC*, JORDON HAUGE, Amalgamated Sugar, 2320 Orchard Dr E, Twin Falls, ID, 83301.

Effects of high silica lime rock on beet sugar operations.

Amalgamated Sugar-Twin Falls sourced a high silica limerock during the length of the 2021 campaign. The silica contained in the rock deposited onto heat transfer surfaces throughout the beet end, mill evaporators and sugar end. Scale formation in beet thick juice evaporator trains has been shown to impact the operation of plate-pack style evaporators most significantly, where scale can potentially block flow of thick juice. Evaporator boilouts were completed using both Hydrite Enhance 567 and Hydri-Maize 2759 to remove scale from the heat transfer surfaces. The use of Hydri-Maize 2759, a peracetic acid-based CIP product has been especially successful in removing large amounts of carbon and silica scale in numerous evaporator effect bodies. Heat transfer calculations across the evaporators were calculated at the tail end of the 2021 campaign and beginning of the 2022 campaign. These calculations point to very limited heat transfer across evaporators due to the silica scale deposited. In addition to evaporator scaling, severe disruptions at the Kelly filter station, extract pan, melters and white pans were most likely caused by silica precipitates. Frequent high-pressure events on the Kelly filters required excessive maintenance to the point of limiting slice rates during campaign. Dark, glass like materials were also observed in the calandria of pans during the sugar end maintenance period. While the mill evaporators suffered the most apparent problems, the effects of the high silica limerock hurt performance throughout the entire plant.

WARDEN, TIM*, Englo, Inc, PO Box 1124, Beckley, WV 25802, One White Oak Trace, Beckley, WV 25801.

The use of wet dust extractors for the prevention of dust explosions in sugar production and handling.

Sugar is combustible and presents an explosion hazard if ignited after it is dispersed as an airborne dust cloud. The handling of sugar fines from crystalline sugar can trigger an explosion. The finer the sugar, the greater the risk of an explosion. Additionally, sugar milling designed to obtain fine particles represents a potential ignition source. In sugar handling and production facilities, fines can accumulate on production floors, along conveyor belts, on machinery, in hot rotary dryers and in steel storage and conditioning silos, in dust collectors, as well as on horizontal surfaces. When sugar is moved or disturbed, it becomes airborne and exposed to oxygen. At that point, an ignition source can result in multiple explosions. Often times a small initial explosion occurs followed by a larger secondary explosion. The first explosion creates pressure waves that can add turbulence and increase dust loading and then create a large ignition source. The National Fire Protection Association (NFPA) codes that apply to sugar processing are NFPA 68 (Explosion Protection by Deflagration Venting); NFPA 69 (Explosion Prevention Systems) and NFPA 654 (Prevention of Fire and Dust Explosions from the Manufacturing, Processing and Handling of Combustible Particulate Solids). NFPA 61 (Prevention of Fires and Dust Explosions in Agricultural and Food Processing Facilities) protects lives and property from fires and dust explosions in facilities handling, processing, or storing bulk agricultural materials, their byproducts, or other agriculture-related dusts and materials. Finally, NFPA 652 (Fundamentals of Combustible Dust) provides the minimum requirements to be met to achieve duct explosion protection and includes the requirement for a Dust Hazard Analysis to be performed. To protect processing equipment and personnel, a multitude of technical measures is often required. They include passive devices (explosion vents), active devices (dust mitigation systems specifically designed to prevent explosions as well as explosion suppression equipment), as well as explosion isolation. Careful attention to prevention, mitigation and isolation will ensure the protection of both personnel and facility. Wet dust extractor systems are designed to prevent combustible dust explosions and are thus safe and reliable. Wet dust extraction systems also typically have a lower initial acquisition cost as well as have a lower operating cost than baghouse type dust collectors.

WARDEN, TIM*, Englo, Inc., PO Box 1124, Beckley, WV 25802, One White Oak Trace, Beckley, WV 25801.

The importance of an efficient flume water filter in ensuring that a beet sugar facility's wastewater treatment solids discharge norms are met.

Flume water filters are used by beet sugar facilities for solid debris removal. Debris and fibers are captured from wash water and are directed onto a filter belt and are then separated with particle size removal flexibility. In some facilities, beet debris containing sugar is recycled. Return on investment from flume water filters can sometimes be achieved in one single sugar campaign. Flume water filters are available in a variety of flow capacities. Units are available having capacities from hundreds to thousands of gallons per minute. Filter belts are constructed of resistant plastic and their replacement, when necessary due to normal wear and tear, can be performed with relative ease. Debris and fibers rest on the surface of the belt and are discarded in the direction that the belt is conveyed for further processing downstream. Filtered water passes through an enclosed sump section of the filter and is then discharged through sump outlets. A flume water filter is a gravity belt filter. It typically consists of a robust frame having a water distribution box on which contaminated water is directed. The flume water is evenly distributed over the entire width of the filter. Flume water passes over a guide plate onto the surface of the belt at the filter overflow. Wash water is filtered through the mesh openings in the filter belt. An efficiently designed and manufactured flume water filter minimizes the solid load to a beet sugar facility's wastewater treatment system, a factor that is critically important for the overall wastewater treatment system performance efficiency. The meeting of discharge norms for solids is in large part dependent on a properly functioning flume water filtration system.

WEBB, DAVID G.*[1], STANLEY D. CASE[1], KRISTOPHER C.C. HENNA[1], MARQUIS G. ATKINSON[1], MICHAEL R. FAGAN[2], JEFFREY W. ZUCEK[2], BRENDA E. BLAKE[2] and MICHAEL L. NELSON[2], [1]Amalgamated Sugar Company, 1951 S. Saturn Way, Suite 100 Boise, ID 83709, [2]USP Technologies, 5640 Cox Rd. Glen Allen, VA 23060.

Increased evaporator performance: Stubborn scale removal using hydrogen peroxide.

Scale deposition in evaporators impedes heat transfer, slows processing, and can lead to degradation of product. Even after standard cleaning methods have been used (caustic wash, back-boiling, mild acid wash, and physical scrubbing), scale can persist and accumulate to make evaporation the rate limiting step of sugar beet processing and can require the replacement of the fouled heat exchange units. Plate pack evaporator fouling has limited mechanical cleaning methods and requires aggressive chemical cleaning methods that are effective in low or no flow conditions. To supplement caustic cleanings, a novel cleaning strategy using hydrogen peroxide in combination with surfactants/solvents was evaluated as a non-capital solution to remove scale buildup inside the evaporators at Amalgamated Sugar's Mini-Cassia factory in 2021 and 2022. Scale loss on ignition analysis and bench scale testing demonstrated localized reaction of hydrogen peroxide. Multiple methods and cleaning procedures were evaluated using concentrations of 7-20% with bursts of 50%; critical operational and safety considerations were addressed prior to implementation. Of seven falling film plate pack evaporators, four were cleaned in 2021 and five were cleaned in 2022. Observations during cleanings (color, flow pattern, steam rates, pressure drop, and visual inspection) indicated tough scale was being removed or dislodged from historically plugged areas inside the plate pack. Over a comparable period, average evaporation rates in 2020, 2021, and 2022 were 579, 625, and 638 tons/hr, respectively; a 10% increase in a two-year period. In 2021, these improvements were used to realize a 6.5% increase in slice rate and in 2022 these improvements are helping to extract a higher percentage of sugar from existing beet crop with high water evaporation rates. Annual proactive cleanings with hydrogen peroxide have optimized traditional cleaning strategies and have yielded incremental improvement to evaporator performance, resulting in increased production. □

POSTER PRESENTATIONS

BOUCHE, CATHERINE*, ITECA, 445 Rue Denis Papin 13592 Aix-en-Provence, France.

Optimizing energy and water efficiency using online pan camera CrystObserver®.

Image processing techniques are now widely applied in many sugar factories willing to optimize process control. With its exceptional video quality and image analysis, the Crystobserver® follows the crystal growth in Real-Time. It detects particles from 4 µm and checks the seeds enter the pan at the right time to avoid uncontrolled nucleation. There is no waste of time or energy before confirming the presence of adequate grain and their development. Any false grain occurring later in the process is detected and water usage is reduced by controlling the remelt. Operators seeing the impact of a parameter modification directly on the crystal size can accordingly establish the best possible sequences to stabilize the massecuite production with a reduced number of fines and better MA and CV. These improvements not only increase pan yield, stability and quality, but they also minimize strike time and costly recycling. They directly translate into benefits from water and energy savings to packing and conditioning enhancements.

BRUNI, GILLIAN O.*, EVAN TERRELL, YUNCI QI and K. THOMAS KLASSON, USDA-ARS Southern Regional Research Center, 1100 Allen Toussaint Blvd., New Orleans, LA 70124.

Characterization of exopolysaccharides from dextran- and fructan-forming microorganisms from sugar crops.

Approximately 8.1 million metric tons of sugar are produced in the U.S. each year with sugar beets and sugarcane contributions at 55% and 45%, respectively. While each of these crops are typically grown in different geographic locations with divergent climates, significant overlap exists in terms of post-harvest processing challenges. Microbes such as *Leuconostoc* that infect sugar beet roots not only result in potential crop loss, but also cause additional problems during post-harvest processing. For instance, microbial degradation of sucrose-rich processing streams results in sucrose losses and exopolysaccharide production that cause economic losses and operational problems during both sugar beet and sugarcane post-harvest processing. Here we isolated and characterized microbial isolates such as *Leuconostoc sp.* from sugarcane post-harvest processing streams that are also present in sugar beet post-harvest processing streams. We show that these strains produce exopolysaccharides in a sucrose-dependent manner. Exopolysaccharides were isolated from microbial culture supernatants and used for structural analysis including Fourier transform infrared spectroscopy (FTIR), nuclear magnetic resonance (NMR), gel permeation chromatography (GPC), and other analytical chemistry techniques that may be useful for determining how composition, molecular weight, and structure may affect viscosity and susceptibility to dextranase treatment during processing. Ongoing work is also exploring qualitative and quantitative techniques to measure the presence of components beyond dextran, including levan fructans and non-sucrose sugars. We are very interested in extending sugarcane analyses to problems relevant for sugar beets, as there is significant overlap in post-harvest processing technologies for the two industries.

KLASSON, K. THOMAS*, GILLIAN O. BRUNI and EVAN TERRELL, USDA-ARS Southern Regional Research Center, 1100 Allen Toussaint Blvd., New Orleans, LA 70124.

Production of chemical precursors from high color sugar beet extract.

Acetoin (3-hydroxybutanone) is a four-carbon ketone-alcohol used in the food industry and is also a precursor to important industrial chemicals such as butanediols and butanols. Acetoin is produced from glucose by the bacterium *Bacillus subtilis*, but the nutrient requirements are unknown when grown on complex sugar sources. Normally, salts, urea, and complex nutrients such as yeast extract and corn steep liquor are added to help the bacterium grow on glucose. We tested high color sugar beet extract as a sugar source for bacterial growth and acetoin production. Based on our initial work, yeast extract can be eliminated without impact on acetoin production. This represents a cost saving to the overall process.

LIMA, ISABEL M.*[1], EMMANUEL SARIR[2], CRAIG PARKER[3], CHARLES CLAYTON[1] and ANNIE TIR[1], [1]USDA-ARS Southern Regional Research Center, 1100 Allen Toussaint Blvd., New Orleans, LA 70124, [2]CarboUA International, Los Angeles, CA, [3]British Sugar, Sugar Way, Peterborough PE2 9AY, UK.

Design and operation of a pilot plant for sugar beet extract colorants removal using powdered activated carbon.

During sugar beet processing, molasses is subjected to slow moving bed chromatography to extract high value betaine and recover more sucrose. The result is commonly a high color sugar beet extract. Sucrose recovery can be increased by recycling this extract from the chromatography system back into the crystallization unit. However, this is only economically viable after it undergoes color reduction. Feasibility pilot plant studies were undertaken on the use of a high surface area powdered activated carbon (PAC) and diatomaceous earth (DE) to adsorb color compounds from sugar beet extract. Color compounds associated with beet extract either are generated during processing or are natural colorants. Experiments were performed using a batch decolorization process to maximize color removal and determine an optimal distribution of PAC either as a body feed or a pre-coat in a filter. A target of 50% color removal was achieved using 4,000 ppm of PAC with a recommended distribution of 75% as pre-coat in the filter and 25% as body feed in the process feed tank. A 50/50 distribution of PAC also produced consistent rate of color removal. PAC performance was slightly better for native colorants to sugar beet than factory colorants. Addition of PAC did not lead to sucrose loss neither had any negative effect on the pH of beet extract.

LIMA, ISABEL M*., USDA-ARS Southern Regional Research Center, 1100 Allen Toussaint Blvd., New Orleans, LA 70124.

Biochars made from pelleted sugar beet pulp for heavy metal remediation.

Processing of sugar beet roots for extraction of white refined sugar also yields sugar beet pulp as a byproduct. Traditionally, sugar beet pulp has been used as animal feed for its nutritional worth but other value-added transformations could further improve the economics of sugar beet production. Thermo-chemical conversion of sugar beet pulp is one such avenue. This process converts the organic carbon rich pulp into a value-added biochar as the main product, but also produces synthesis gas and bio-oil, two products that could fit into alternative energy frameworks. This study looks at the physico-chemical and adsorptive properties of pelleted sugar beet pulp biochars produced at four different pyrolysis temperatures: 350, 500, 650 and 800°C for 1 hour residence time and investigated various uses from adsorption/remediation materials to source of fuel and use as soil amendment. Biochars were able to adsorb various heavy metals from solution, particularly lead, chromium and copper. While lower pyrolysis temperature biochars performed better in adsorption experiments, higher temperature biochars were better candidates for fuel applications. Due to their N-P-K composition, biochars from sugar beets also showed potential to be used as soil amendments to enhance soil health and improve crop yields.

QI, YUNCI* and GILLIAN O. BRUNI, USDA-ARS Southern Regional Research Center, 1100 Allen Toussaint Blvd., New Orleans, LA 70124.

Genomic analysis of polysaccharide-producing bacteria from sugarcane processing streams.

Microbes cause economic losses during postharvest processing of both sugar beet and sugarcane by consuming sucrose and producing extracellular polysaccharides (EPS), which negatively affect the purification and crystallization process. Recent research has uncovered a wide diversity of sugar-consuming microbes present in sugar factories, and the structure and composition of the EPS they produce are likely just as diverse. An improved understanding of the microbes in sugar factories will help guide improvements in measures to inhibit microbial growth and remove EPS through the use of hydrolyzing enzymes. Here, we report on the whole-genome sequencing of 24 bacterial strains recently isolated from sugarcane processing streams in Louisiana, which include *Leuconostoc, Gluconobacter,* and *Pantoea* species. We were able to identify the EPS-producing enzymes encoded on each genome, which varied in both function and copy number per genome. Finally, the isolated bacteria were spotted on various sugar substrates, revealing a diversity of EPS-producing phenotypes that reflect their genetic diversity. We show that genomics is a powerful tool in predicting the structure and composition of EPS produced by microbes present in sugar factories, which will help guide improvements in enzymatic removal of EPS.

PHILIP SCHOFIELD* and MARKUS WEINGARTNER, ANDRITZ Fiedler GmbH, Weidener Str 9, 93057 Regensburg, Germany.

Improvement of the efficiency of a batch centrifugal by basket replacement.

Sugar producers are looking to improve the efficiency of the centrifugals in their process. ANDRITZ was asked whether the efficiency of the batch centrifugal basket could be improved, so that when batch baskets are replaced, a process efficiency could be achieved. We used in house engineering and filtration expertise to develop a replacement batch centrifugal basket that offers shorter cycle times, increased sugar output per cycle, improvement in sugar moisture and longer life of wedge wire inner screens in a lighter, long-life basket that delivered better efficiencies and met the expectations of the sugar producers. ☐

NOTES

American Society of Sugar Beet Technologists
800 Grant Street, Suite 300
Denver, Colorado 80203-2944

ISBN 978-1-7138-7325-9

CURRAN ASSOCIATES INC.
proceedings.com